King's College London

The Maughan Library & Information Services Centre
Chancery Lane
London WC2A 1LR

Telephone: 020 7848 2424

This book may be recalled at any time and must be returned or
renewed by the date shown.

Environmental Policy Integration in Practice

Environmental Policy Integration in Practice

Shaping Institutions for Learning

Edited by

Måns Nilsson and Katarina Eckerberg

London • Sterling, VA

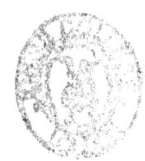
First published by Earthscan in the UK and USA in 2007

ISBN: 978-1-84407-393-1
Typeset by Composition and Design Services
Printed and bound in the UK by Cromwell Press, Trowbridge
Cover design by Susanne Harris

For a full list of publications please contact:
Earthscan
8–12 Camden High Street
London, NW1 0JH, UK
Tel: +44 (0)20 7387 8558
Fax: +44 (0)20 7387 8998
Email: earthinfo@earthscan.co.uk
Web: **www.earthscan.co.uk**

22883 Quicksilver Drive, Sterling, VA 20166-2012, USA

Earthscan is an imprint of James and James (Science Publishers) Ltd and publishes in association
with the International Institute for Environment and Development

A catalogue record for this book is available from the British Library

Library of Congress Cataloging-in-Publication Data
Environmental policy integration in practice : shaping institutions for learning / edited by Måns
Nilsson and Katarina Eckerberg.
    p. cm.
 ISBN-13: 978-1-84407-393-1 (hardback)
 ISBN-10: 1-84407-393-9 (hardback)
 1. Environmental policy--Sweden. 2. Environmental protection--Sweden. I. Nilsson, Måns,
1969- II. Eckerberg, Katarina, 1953-
 GE190.S8E68 2007
 333.7209485--dc22
                        2006039462

The paper used for this book is FSC-certified and
totally chlorine-free. FSC (the Forest Stewardship
Council) is an international network to promote
responsible management of the world's forests.

Mixed Sources
Product group from well-managed
forests and other controlled sources
www.fsc.org Cert no. TT-TOC-2082
© 1996 Forest Stewardship Council

Contents

List of Figures and Tables

Figures

Tables

About the Contributors

Editors

Måns Nilsson is a Research Fellow at the Stockholm Environment Institute (SEI) and Director of the SEI's Policy and Institutions Programme. He specializes in environmental policy analysis, institutional development and strategic assessment, with a particular emphasis on climate and energy policy and development policy. He received his PhD degree in policy analysis from the Delft University of Technology. In 2001–2002, he was a fellow with the Carnegie Council for Ethics and International Affairs. He has published widely in academic journals related to policy analysis, assessment, energy and environmental policy. Prior to joining the SEI, he worked with UNDP/GEF in New York and Vattenfall in Sweden.

Katarina Eckerberg is Professor in Public Administration at the Department of Political Science, University of Umeå, specializing in environmental politics. In 2006 she joined the SEI as Deputy Director. Her research and teaching focuses on the implementation of sustainable development policy, land use policy and management of natural resources, particularly in Sweden, with comparisons across Europe and involving issues of multi-level governance from local to global. She was Visiting Scholar at the University of Cambridge and the London School of Economics and Political Science in the 1990s and worked with the FAO Community Forestry Programme at the Policy and Planning Unit in Rome in the late 1980s.

Contributing Authors

Rebecka Engström has a PhD degree from the Division of Environmental Strategies Research at KTH (the Royal Institute of Technology). Her research explores strategies for reduced environmental impacts by using methods of environmental systems analysis such as input–output analysis and life-cycle assessment. Her previous main interest lay in strategies for reduced environmental impacts from food production and consumption.

Lovisa Hagberg works at the Swedish National Board of Forestry. She received her PhD from the Department of Political Science, University of Umeå. Her thesis concerned the consequences of globalization for the ways in which political space-time has been theorized in green political thought. Previously she worked with questions of implementation in agri-environmental policy. She has taught human ecology and, with her background in forestry, has an interest in interdisciplinary studies.

Göran Finnveden is Associate Professor in Industrial Ecology and Head of the Division of Environmental Strategies Research at KTH (the Royal Institute of Technology). He specializes in developing and using tools for environmental assessment of products, systems, policies, plans and sectors. He received his PhD in natural resources management from the Department of Systems Ecology, Stockholm University. He has published almost 40 papers in scientific journals and is on the editorial board of several scientific journals.

Åsa Persson is a Research Associate at the Stockholm Environment Institute and holds a PhD degree from the London School of Economics and Political Science. Her research interests are the process of choosing environmental policy instruments within the waste management sector, environmental policy integration, and evaluation of environmental policy. She joined the SEI in 2001 and has since published her work both in academic journals and policy-oriented reports.

Åsa Gerger Swartling is a Research Fellow at the SEI. She specializes in participatory approaches to environmental policy and management, sociology of science, and policy analysis. She received her PhD degree in sociology from the University of York. Her fieldwork has included countries in the North (Sweden, the UK and EU-level), the South (Latin America, the Caribbean and South and Southeast Asia) and Eastern Europe.

Charlotta Söderberg is a PhD candidate at the Department of Political Science, Umeå University. She specializes in environmental policy integration in Swedish bioenergy policy. She has a MSc in political science from Luleå University of Technology, where she wrote her thesis on policy options for introducing environmental taxation on virgin materials and chemicals in Sweden.

Acknowledgements

The material presented in this book is based on studies carried out in the research project 'Policy Integration for Sustainability', funded by the Swedish Research Council Formas from 2002 to 2006. We, along with all the other chapter contributors, are grateful for the generous financial support provided through this grant.

The work has benefited immensely from in-depth consultations and many challenging discussions with our esteemed scientific and policy advisers. A warm thank you goes out to Roger Kasperson, Wil Thissen, Stig Wandén, Olle Björk, Annika Löfgren, Ingrid Swedinger, Andrew Jordan, Andrea Lenschow, Mikael Hildén, Yvonne Rydin, Lewis Gilbert and Malur Bhagavan.

We would also like to warmly thank the vast number of people (who remain anonymous) in and around the policymaking arena in Sweden who openly and candidly shared with us their insights and experiences. The study would not be worth anything without your input.

Finally, thank you to our co-authors Charlotta Söderberg, Lovisa Hagberg, Åsa Persson, Åsa Swartling, Rebecka Engström and Göran Finnveden for your hard work preparing the chapters and contributions to the whole idea and concept behind the book.

Måns Nilsson and Katarina Eckerberg
Stockholm
September 2006

Acronyms and Abbreviations

CAP	Common Agricultural Policy
CHP	combined heat and power
CO_2	carbon dioxide
EAP	environmental action programme
EC	European Commission
EEA	European Environment Agency
EMS	environmental management system
EPI	environmental policy integration
ERDP	Environmental and Rural Development Plan
EU	European Union
EU-15	European Union's 15 member states (1995–2004)
GATT	General Agreement on Tariffs and Trade
GMO	genetically modified organism
HEPI	horizontal environmental policy integration
IOA	input–output analysis
IPCC	Intergovernmental Panel on Climate Change
LAG	parliamentary working group for agricultural reform
LCA	life-cycle analysis
LO	Swedish Trade Union Confederation
LRF	Lantmännens Riksförbund (Federation of Swedish Farmers)
MFA	material flows analysis
MoA	Ministry for Agriculture
MoE	Ministry for the Environment
MoF	Ministry of Finance
MoI	Ministry for Industry and Trade
NBA	National Board of Agriculture
NEQO	national environmental quality objectives
NGO	non-governmental organization
NUTEK	Swedish Agency for Economic and Regional Growth
OECD	Organisation for Economic Co-operation and Development
Prop	Proposition (government bill)
R&D	research and development
RCT	Rational Choice Theory
REAS	Rural Economy and Agricultural Societies
Rskr	Riksdagens Skrivelse (parliamentary communication)
SCB	Statistiska Centralbyrån (Statistics Sweden)
SDS	sustainable development strategy

SEA	strategic environmental assessment
SEPA	Swedish Environmental Protection Agency
SFIF	Swedish Forest Industries Federation
Skr	Regeringens Skrivelse (government communication)
SOU	Statens Offentliga Utredningar (official committee report)
SSNC	Swedish Society for Nature Conservation
STEM	Statens Energimyndighet (Swedish Energy Agency)
SUAS	Swedish University of Agricultural Sciences
Svebio	Swedish Bioenergy Agency
UNCED	United Nations Conference on the Environment and Development
VEPI	vertical environmental policy integration
WCED	World Commission on Environment and Development
WTO	World Trade Organization
WWF	World Wide Fund for Nature

Forewords

I

In 1987 the Science Advisory Board of the US Environmental Protection Agency published an influential report, *Unfinished Business* (US EPA, 1987), which gauged the success of environmental regulation in the US since the founding of the Environmental Protection Agency in 1972. It followed this with *Reducing Risk* (US EPA, 1990). The message was sobering. The easiest part of the job had been done, according to the report, but the hard work was yet to occur. Past efforts had focused on 'end-of-the-pipe' solutions; the next generation efforts would need to move upstream to address the driving forces of environmental change and environmental risks. This meant, in practice, intervening into fundamental sectors of the economy – energy, transportation, agriculture. These are very much the issues that this volume addresses in a Swedish context. The report predicted that the 'second generation' environmental efforts would be far more challenging than the 'first'. They would need to become mainstream interventions – policy would need to be integrated into the overall management of industrial sectors and management would need to focus much more on policy implementation than policymaking alone. The book that follows assesses the progress and challenges in Sweden, a country on the forefront of an historic transition.

All this anticipates the metamorphosis of environmental protection policy frames to sustainability thinking. The focus shifts from cleaning up the 'externalities' of industrial processes to the mission of creating a more sustainable society. This is a process whose face extends horizontally across governmental agencies and the private sector, but also vertically into societal values and institutions. This explains why the authors of this volume, and others, argue that the battles in this transition are not to be won overnight. Fundamental change in basic economic and societal structures comes slowly, to say nothing of transformation of institutions and societal values. The authors of this book probe how change may occur, wedding conceptual insights with empirical results based on significant sectoral changes over the past 15 years.

The evolving context of political culture and social awareness cannot be underestimated as a fulcrum of what can be accomplished. Social transformation is embedded in the quest for policy integration and, more ambitiously, for a sustainable society; without it progress may be more illusory than real, more ephemeral than durable, more partial than holistic. So it must be appreciated that the past several decades have been a fertile ground for environmental policy integration – advanced industrial societies have become more risk averse (what

Beck (1992) called the 'risk society'), attitudes toward technology have increasingly shifted from the assumption of progress to that of threat, expectations for consultation and participation have increased dramatically, and environmental values remain durable in the face of high unemployment rates and rising energy prices. To what extent continuity can be expected in these trends and the favourable setting for sustainability initiatives is an open question.

So understanding the process by which environmental policy integration may proceed is essential to assessing the prospects for future change. The volume that follows, drawing upon such seminal works as Argyris and Schön (1978), Sabatier and Jenkins-Smith (1993) and Jasanoff (1999), gives particular focus to social learning as a means of policy integration. This is an important line of inquiry, although there is much to be learned empirically about the different types of social learning that may occur and how instrumental they are in policy integration. Social networks are likely to be very important in these integrative processes, both the integration that extends beyond institutions and sectors or policy domains but also those that link the sources of new knowledge and expertise with decision makers but also decision makers and their clients. Here there may be a role also for exploring not only the structures of these social networks but also diffusion theory that treats policy integration as an innovation that spreads across institutional actors, with early and late adopters, and what determines the speed of diffusion and the impediments that emerge.

One issue that this study provokes is the relation between policy integration across sectors and the possibilities for a shift from command-and-control cultures to more adaptive management strategies. Interest in adaptive management and decision making under uncertainty is growing rapidly, and appropriately so. Does greater policy integration facilitate or impede the movement to more adaptive management strategies? There is much to be learned here and this volume is a good point of embarkation.

Roger Kasperson
Washington, DC
August 2006

Roger E. Kasperson is Professor of Government and Geography at Clark University. Dr Kasperson is a member of the US Academy of Sciences and serves on the Executive Committee of the US Environmental Protection Agency's Science Advisory Board. He received his PhD degree from the University of Chicago.

II

Environmental policy integration (EPI) was one of the most powerful concepts to emerge in environmental policy discourse in the late 20th century; others, closely related, were sustainable development and ecological modernization. Throughout the Western world, significant institutional change and policy reform took place after the environmental revolution of the 1960s and early 1970s. By the 1980s, however, it was becoming apparent that environmental protection would always be reactive, and probably inadequate, if it was treated as a postscript to policies in other, established sectors. Thus the integration of an environmental dimension into key policy domains like energy, agriculture and transport has become one of the central tenets of ecological modernization, and is widely regarded as a prerequisite for sustainable development.

EPI has strong normative dimensions: we conceive of it in terms of what *ought* to happen at the interface of society, economy and environment. As with other important concepts in human affairs, however, it is underdetermined by its exposition in principle. 'Usable doctrines', as David Marquand (1988, p12) has observed, 'have to be hammered out in the give and take of a debate, provoked and shaped by the lived experience of particular societies at particular times.' Not surprisingly, then, EPI is contested and refined as we seek to interpret and implement it in practice. One of the strengths of this book is that it approaches its subject from both theoretical and empirical perspectives, providing on the one hand a conceptual and political history of EPI and on the other a detailed account of changing practices and outcomes in the agricultural and energy sectors in Sweden. A Swedish analysis is of interest because, to paraphrase the authors themselves, if EPI doesn't work in Sweden, it won't work anywhere. It is also good to see theory and empirical work being employed together to advance our understanding of societal and environmental change: in the literature on environmental issues and policies, the theoretical and the empirical have all too often been divorced.

In the broadest sense, one might argue that a degree of integration is already apparent in advanced industrial economies. Few would deny that the environmental implications of human activities are subject to much greater scrutiny than they were 30 or 40 years ago. This is due in part to formal processes of environmental assessment, which have contributed in a modest way to the greening of growth and, perhaps more important, have offered opportunities for learning within and between policy communities (Owens et al, 2004). Greater public and political awareness, and rafts of legislation, have also played their part. Yet in spite of these developments, there remains a sense of implementation deficit about EPI. As the authors note, the so-called 'Cardiff Process', which in the late 1990s held out such promise for EPI at the European level, has apparently foundered; and their own detailed analysis concludes that integration in the context of Swedish agricultural and energy policies as been partial, uneven and contextually specific.

Why is genuine integration so difficult? One answer – the most encouraging – may be that we are looking for fundamental changes too soon. Ideas, it is now widely acknowledged, have an impact on policy, but the process can take many years. EPI did not emerge as a fully fledged idea until the 1980s, and to be effective it must impinge on sectors with economic, cultural, political and strategic significance. We should not be surprised, therefore, if the process of transformation is incomplete, and perhaps we just need to be more patient. But there is another, and less comfortable, interpretation. Integration is difficult because it forces us to make choices. Having made the 'easy gains', we must now confront deeply embedded sectoral objectives if we are not to breach environmental capacities. The UK energy sector (of which I have some experience in my own research) illustrates this line of thought. We tackled the most obvious impacts (urban air pollution, acid rain, radioactive discharges) in the 20th century, and became noticeably more energy efficient. But in the 21st century we face new challenges – and new manifestations of old conflicts. We must deal with climate change – a less tangible and more ubiquitous problem than the gross pollution of 50 years ago. At the same time we are taking up efficiency gains in increasingly extravagant patterns of consumption (think of large vehicles, cheap flights and domestic appliances). On the supply side, the nuclear option remains controversial, and we are finding that 'renewable' is not synonymous with 'environmentally benign'. In such circumstances, as the authors of this volume recognize, integrating an environmental dimension into energy policy can expose, rather than reconcile, fundamental conflicts of interest and value.

Nevertheless, we can and should derive positive messages from this work. I believe that the authors are right to conceptualize EPI in terms of a process of social learning, in which cognitive factors, including ideas, argument and analysis, are seen to have independent significance in the policy process. Of course learning will always take place in interplay with interests and power. Crucially, though, as many authors have recognized, interests may not be fixed, but can themselves become 'a dynamic "dependent" variable, framed by knowledge' (Radaelli, 1995, p165). The reconstruction of interests is especially likely in situations characterized by uncertainty, and it has important implications for policy change. For a number of reasons, therefore, I warmly welcome this book. It provides not only a substantive analysis of the theory and practice of EPI, but case studies of policy learning. In doing so it offers valuable new insights into both.

Susan Owens
Cambridge
August 2006

Susan Owens is Professor of Environment and Policy at the University of Cambridge. She has been a standing member of the Royal Commission on Environmental Pollution since 1998 and has served on a number of other public bodies. In 2000 she received the Royal Geographical Society's 'Back' Award. Her PhD is from the University of East Anglia.

References

Argyris, C. and D. Schön (1978) *Organizational Learning: A Theory of Action perspective*, Addison-Wesley, Reading, UK
Beck, U. (1992) *Risk Society: Towards a New Modernity*, Sage, London
Jasanoff, S. (1990) *The Fifth Branch*, Harvard University Press, Cambridge, MA

Marquand, D. (1988) *The Unprincipled Society: New Demands and Old Politics*, Fontana Press, London

Owens, S., Rayner, T. and Bina, O. (2004) 'New agendas for appraisal: reflections on theory, practice and research', *Environment and Planning A*, vol 36, no 11, pp1943–1959

Radaelli, C. M. (1995) 'The role of knowledge in the policy process', *Journal of European Public Policy*, vol 2, no 2, pp159–183

Sabatier, P. and Jenkins-Smith, H. (eds) (1993) *Policy Change and Learning: An Advocacy Coalition Approach*, Westview, Boulder, CO

US EPA (1987) *Unfinished Business: A Comparative Assessment of Environmental Problems*, United States Environmental Protection Agency, Washington, DC

US EPA (1990) *Reducing Risk: Setting Priorities and Strategies for Environmental Protection*, United States Environmental Protection Agency, Washington, DC

1
Introduction:
EPI Agendas and Policy Responses

Måns Nilsson, Katarina Eckerberg and Åsa Persson

We know more and more about the environmental implications of human activities and what we must do to remedy them. This has fuelled remarkable advancements in environmental policy during the last three decades. However, whether these insights result in modified decisions in mainstream economic policy sectors such as agriculture, energy or transport is another matter. Indeed, sectoral policymaking, where most important public policy decisions are made, appears to remain largely business-as-usual, leaving sustainability concerns marginalized. This book tackles this critical policy problem, persistent across the world: the concern that environmental values are not sufficiently integrated into mainstream sectoral policymaking.

This concern has spread, with the result that integration of environmental sustainability considerations into social and economic policies, the so-called environmental policy integration (EPI) principle, has become widely identified as a necessary shift in policymaking. EPI is deemed an essential component of sustainable development, as we move away from end-of-pipe solutions to proactive solutions to environmental problems. Only by strategically involving all sectors can the momentum be raised to push the whole of society in a more sustainable direction. This necessity of considering economic and environmental policy together has been widely recognized in recent decades and is emphasized in several classical texts. For instance, in 1987 the World Commission on Environment and Development (the Brundtland Commission) argued that:

> *The integrated and interdependent nature of the new challenges and issues today contrasts sharply with the nature of the institutions that exist today. These institutions tend to be independent, fragmented, and working to relatively narrow mandates with closed decision processes. Those responsible for managing natural resources and protecting the environment are institutionally separated from those responsible for managing the economy. The real world of interlocked economic and ecological systems will not change; the policies and institutions must.*
> (WCED, 1987, p310)

In Europe, EPI now has a constitutional backing not only in many national juris-dictions but also in the European Treaty. The Swedish Constitution has a formula-tion that 'public activity shall promote sustainable development leading to a good environment for present and future generations' and Article 6 of the European Treaty of 1997 ('The Amsterdam Treaty') declares that 'Environmental protec-tion requirements must be integrated into the definition and implementation of the Community policies [...] in particular with a view to promoting sustainable development.'

Hence the principles, policies and declarations regarding EPI have been in place for many years. But are they followed through in practical terms? As noted above, most observers suggest that the degree of implementation of EPI has been disappointing so far and requires more attention, both in practical policymaking and in policy-oriented research (Lenschow, 2002a; EEA, 2005). Just as with sus-tainable development, the idea of EPI has largely remained on the rhetorical level (EC, 2004). Why, despite political and constitutional support, is EPI largely con-sidered a failure internationally? How do we really measure success when it comes to EPI? What are the barriers and challenges and how can they be overcome? What experiences can we draw upon to enhance the potential for EPI?

Building on national policy experiences in Sweden, this book is an attempt to unpack these questions, putting two overarching issues up front: First, how can one meaningfully interpret the principle of EPI, and what is its status and development in policy sectors? Second, what main factors affect EPI and how should institutions in the policymaking arena be shaped to advance it? The book presents a multidisciplinary study of these questions, focusing on empirical cases in the agricultural and energy policy sectors in Sweden. These sectors are interest-ing because they are at the heart of some of society's major sustainability chal-lenges, but represent very different types of issues. In addition, they have recently been subject to intense EPI efforts in many countries. The book uses sector envi-ronmental analysis, policy framing and learning, and the role of institutions to unpack the EPI puzzle and its evolution in these sectors. With a focus on institu-tional dimensions, we intend to provide empirical insights and recommendations to practitioners engaged in EPI efforts across Europe and elsewhere; we also hope to provide valuable theoretical insights for students of environmental policy.

Why is the EPI experience in Sweden interesting as a case study? Among both Swedish policymakers and those abroad, Sweden is perceived as a pioneer in envi-ronmental policy (Skou Andersen and Liefferink, 1997; Lafferty and Eckerberg, 1998) and has also been known to push the environmental agenda in interna-tional arenas such as the EU and the UN, including the EU's sustainable develop-ment strategy (SDS) and the Cardiff Process. Its track record in environmental policy performance consistently comes out high in international comparisons. For instance, *Time Magazine* featured an article on Sweden as *the* main front-runner country in the world:

> *Sweden's leaders have passed laws that would be unthinkable for some politicians [...] Indeed, whereas others complain about higher taxes or infringements on their rights, most Swedes seem to embrace the idea of helping save the planet.* (*Time Magazine*, 2006, p41)

At the World Economic Forum in 2006, Yale University presented a study assessing the environmental performance of 133 countries. According to Yale's 2006 Environmental Performance Index, Sweden ranks second after New Zealand (and followed by Finland, the Czech Republic, the UK, Austria and Denmark) (Yale Center for Environmental Law and Policy, 2006). Because Sweden is so successful at environmental policy we expect it to be good at EPI as well, and the study of Sweden can be expected to also generate lessons for other countries.

In Sweden, as elsewhere, many actors and institutions involved in policymaking lack sustainability perspectives through tradition and mandate. Sweden has certainly experienced its share of 'environment-bashing' and vested interests stalling progress towards sustainability. Nevertheless, it is apparent that some things work well in Sweden when it comes to environmental policy. And as we shall see later on, it has responded actively to the challenges set by the EPI agenda and the Swedish policy system displays several features such as consultative, coordinative and fact-finding processes that appear to be beneficial for EPI. Although research has shown that there is not one governance structure or institutional set-up that works best for environmental decision making everywhere (Brewer and Stern, 2005), the principles and mechanisms behind the Swedish arrangements should be broadly interesting to learn from. However, before looking more closely at the Swedish response, we will briefly introduce the EPI concept and the key international and European processes that contributed to establishing the EPI agenda.

EPI with a View Towards Sustainable Development

What does EPI really mean? EPI can be distinguished from environmental policy and protection in general in that it is concerned with environmental policy issues becoming part of sectors rather than being a separate (albeit sometimes successful) policy field on its own. However, beyond this there is little agreement on the exact meaning of the concept, as explored in detail in Chapter 2 and also discussed in earlier literature (Lafferty, 2004; Lenschow, 2002a; Jordan and Schout, 2006). The principle, as expressed in the European Treaty and in the Brundtland Report, seems straightforward, suggesting that environmental issues need more consideration within sectors. How much consideration is, however, a point of contention, with some authors going as far as to suggest a general priority for the environment over other objectives as a general rule of EPI (Lafferty and Hovden, 2003) and others seeing it as a way of establishing a more rational policymaking process by ensuring a more comprehensive basis for decisions (Underdal, 1980) (see Chapter 2 for an in-depth discussion).

Ultimately, the process of integration is a context-specific interpretation process which is likely to involve many different actors and evolve over time as problems and understandings are continually reframed. This means that for the purposes of this book, specific operational definitions of sustainability cannot be theoretically deduced and used as benchmarks. In our research we will refrain from viewing certain measures, technologies or policy agendas as inherently sustainable. Viewing sustainability as a matter of context-specificity and interpretation is consistent with the policy-learning approach to EPI that we apply in this book. The policy-learning approach to EPI tries to bridge the rational and normative perspectives and situate itself somewhere in the middle. We recognize that EPI is something more than rational decision making and entails a normative orientation. Still, since environmental concerns in sectoral policymaking are systematically undervalued, despite democratically established policy goals about environmental protection and sustainable development, a more rational pursuit of our common goals requires enhanced attention to environmental knowledge, at least in the major economic sectors. Thus a policy-learning process drives a normative reorientation in the sectors as well as rendering it more rational and effective vis-à-vis established policy objectives. The learning concept applied in this book does not, however, entail any general priority of environmental objectives over other policy objectives (see Chapters 2 and 3). At some point there will be 'enough' EPI. However, taking a (learning) process view of EPI, we refrain from defining such an end state. The learning conceptualization will be further discussed in Chapter 3 and then applied in the empirical Chapters 5 and 6.

This book is primarily concerned with the environmental objectives of EPI. However, since EPI as a policy principle both in Sweden and internationally is explicitly concerned with integration of the environment into decision making with a view to promoting sustainable development, it must be recognized that EPI in a sustainable development context entails more than protecting the environment in a narrow sense. From an empirical point of view, it is difficult to work with the sustainability concept entailing the environmental, economic and social dimensions that were outlined in the Brundtland process. These three cover virtually all public policy areas and objectives, including employment, work safety, health and industrial competitiveness. Integrating sustainable development would then be a matter of 'integrating everything into everything'. How, then, can 'in the context of sustainable development' be interpreted? The Brundtland definition – 'sustainable development is development that meets the needs of the present without compromising the ability of future generations to meet their own needs' – gives three useful leads. First, being concerned with future generations means that we are concerned about effects far into the future. This means that policy must take a long-term view and potentially trade off goals over time. Second, meeting the needs of the present stipulates efficiency in the use of resources. This means that environmental policy must be efficient in terms of how much resources we spend, or opportunities we give up, to protect the environment. And third, being concerned with the needs of the present also means that policy must take an international perspective and see national policy in a global and equity context.

International Policy Background

The recent impetus to pursue EPI more actively in many countries and international organizations builds on a policy debate with a long history. We have already quoted the Brundtland Commission, which also stated that the consideration of ecological dimensions at the same time as other dimensions is the 'chief institutional challenge of the 1990s' (WCED, 1987, p313). As a result, a number of proposals for institutional change were made. At the national level, it was proposed that:

> *Sustainable development objectives should be incorporated in the terms of reference of those cabinet and legislative committees dealing with national economic policy and planning as well as those dealing with key sectoral and international policies. As an extension of this, the major central economic and sectoral agencies of governments should now be made directly responsible and fully accountable for ensuring that their policies, programmes and budgets support development that is ecologically as well as economically sustainable.* (WCED, 1987, p314)

In 1992 the UN Conference on Environment and Development (UNCED), held in Rio de Janeiro, picked up the views expressed by the Brundtland Commission, devoting Chapter Eight of Agenda 21 to the integration of environment and development in decision making (UNCED, 1992). It identified four programme areas: integrating environment and development at the policy, planning and management levels; providing an effective legal and regulatory framework; making effective use of economic instruments and market and other incentives; and establishing systems for integrated environmental and economic accounting. Similar to other political declarations and strategies at the global level, Agenda 21 lacks more concrete measures, practical advice and timetables for commitments to substantive environmental outcomes to be achieved. All such measures are left to the implementation at national or sub-national levels. The strategy therefore depends on the interest and resources of the individual countries.

The applied work of the Organisation for Economic Co-operation and Development (OECD) on policy approaches towards sustainable development has also contributed to EPI at the international level. One major effort is the initiative to develop indicators to measure the relationship between the environment and sector activities. Several sets of sectoral environmental indicators are in place, including within the transport and agriculture sectors (OECD, 1997 and 1999), which are regularly reported on. Also the OECD's Environmental Performance Reviews regularly take stock of environmental policy and EPI. In addition to policy advice and guidance for sustainable development, the Public Management Service of the OECD has provided advice on policy coherence and integration in general (OECD, 1996). Similar to Agenda 21, the work of the OECD is process oriented and leaves substantive commitments to be made by national decision makers. However, it is commonly regarded as an effective forum for policy diffusion, learning and transfer of experience and best practice between countries.

The European Union and the Cardiff Process

As opposed to the work of the UN and OECD, the European Union (EU) has its own political mandate for environmental policy which means that it can both apply EPI in its own legislation and operation and push member states to apply it in policy areas not controlled by EU legislation. The history of EPI progress within the EU has interested many researchers and is well documented (Weale and

Williams, 1993; Liberatore, 1997; Lenschow and Zito, 1998; Lenschow, 2002b; Nilsson and Persson, 2003). The need for environmental integration in EU sectoral policies was articulated in the Third Environmental Action Programme (EAP) of 1983 and subsequently elaborated in both the Fourth and Fifth EAPs. It was given legal status in the Single European Act of 1987 (Article 130r).

According to Liberatore (1997) several factors contributed to the emergence of the integration principle, including concerns over the environmental implications of completing the internal market, the increasing influence of the discourse of sustainable development, integration initiatives by some member states and the ongoing process of strengthening EU institutions for the environment. However, despite the early attention, the implementation of EPI up to 1992 was 'a faltering and haphazard affair' (Weale and Williams, 1993, p49). For example, environmental aspects were not important decision premises in the creation of the single market. In 1992 the Maastricht Treaty gave EPI a more prominent legal basis. The constitutional basis was completed with the Amsterdam Treaty, in which Article 6 (see above) came into force. However, as expressed in the EC's Second Assessment of 1999 (EEA, 1999), 'the commitment by other sectors and by member states to the programme is partial [...] without a reinforced integration of environmental concerns into economic sectors [...] our development will remain environmentally unsustainable overall.'

In 1998 the European Council initiated the Cardiff Process to achieve more effective EPI in sectoral policymaking. Driven by three new member states – Sweden, Austria and Finland – the Council decided that its configurations should adopt their own strategies for integrating environment and sustainable development (EC, 1998). Apart from environmental protection concerns, a key motivation behind this initiative was the general acknowledgement that the segmented and hierarchical EU institutions produced incoherent policies (Hertin and Berkhout, 2001). The strategies were to be regularly monitored and evaluated with the help of fixed timetables and indicators. The first phase developed strategies and indicators in the transport, energy and agriculture sectors, involving both the European Council and the respective directorates of the European Commisson (EC). This was to be followed by development aid, the internal market, industry and enterprise, general affairs, economics and finance, and fisheries.

Both external evaluations and internal stocktakings of the Cardiff Process show that progress varied considerably between different sectors (Kraemer, 2001; Fergusson et al, 2001; EC, 2004). Some strategies build on incomplete or absent sector assessments, went through little external consultation, and consisted of little more than recapitulation of existing measures. Until 2001 progress had been uneven, with a general failure to comply with the timetable (Fergusson et al, 2001). The Swedish EU presidency in 2001 concluded that the sector work should have a more comprehensive approach, involving in particular the SDS (see below), the Sixth EAP and future reviews of the common policies. Furthermore, gaps in the information systems were identified, including a general assessment of the effects of the strategies in implementing Article 6. As we write this book (in 2006), most observers consider the Cardiff Process to be effectively dead (Jordan and Schout, 2006).

Under the Swedish presidency in 2001, the Council adopted a European 'Strategy for Sustainable Development' (SDS) in Gothenburg (EC, 2001). The strategy is actually an addition of the environmental dimension of sustainable development to the Lisbon Strategy for meeting economic and social objectives. The strategy identifies four priorities: limiting climate change and increasing the use of clean energy; addressing threats to public health; managing natural resources more responsibly and improving the transport system. EPI, or rather 'sustainability policy integration' in this case, has a central role in the strategy, meaning that a broader set of social and environmental concerns need to be considered in decision making in all sectors. It is stated that 'too often, action to achieve objectives in one policy area hinders progress in another' and that 'the absence of a coherent long-term perspective means that there is too much focus on short-term costs and too little focus on the prospect of longer term "win–win" situations' (EC, 2001, p5). After this problem identification, the strategy boldly states that 'sustainable development should become the central objective of all sectors and policies' (EC, 2001, p6).

These statements reflect a relatively strong commitment to fundamental changes in policymaking within the EU. It is fair to say that through these foundations in the EU, the political legitimacy of EPI has been clearly established (Lafferty and Hovden, 2003). However, the actual commitment to implementing and applying the SDS in real policymaking has been questioned, and its status vis-à-vis other processes such as the Lisbon Agenda has proven to be rather weak. The fact that it took 15 years from putting EPI on the agenda to taking real action within the sectors points to possibly strong resistance in the policy mainstream.

Sweden was a main force behind both the Cardiff Process in 1998 and the SDS in 2001. In the next section we will see how Sweden has pursued the principle of EPI in domestic policymaking, with a range of EPI measures implemented throughout the 1990s and early 2000s.

The Swedish Response to the EPI Agenda

The Swedish EPI response has its roots in a long history of environmental protection. Sweden experienced parallel developments of environmentalism, economic development and environmental policy throughout the 20th century, with environmental historians distinguishing between blue and green lines of environmental protection (Lundgren, 1995; Vedung, 1991). The green line was concerned with conservation of nature, first with the objective of conserving species, habitats and environments for allowing scientific inquiries and analyses, but over the years broadened to include concerns for cultural and social values as well as the preservation of landscapes. Within the green line, the Swedish Society for Nature Conservation (SSNC), the largest environmental non-governmental organization (NGO) in Sweden, was formed back in 1909, which was also the year of the first nature protection law. The blue line was concerned with controlling pollution and degradation. This line of policy started from an increasing recognition of the public health problems in urban areas as a result of rapid industrialization in the 19th century. In those days, however, technical experts thought that dilution was enough to address the environmental problems. Until the 1960s environmental protection was a marginal policy concern, but by this time public and political awareness rose rapidly. The number of environmental bills doubled from 1965 to 1968 (Löfstedt, 2003), with the environmental protection act of 1967 providing, for the first time, a holistic perspective that pulled the two lines together.

From the late 1960s, environmental policy in Sweden was characterized by a regulatory, end-of-pipe approach in which 'interest balance' between environmental and other societal interests was a key theme. Up until the 1980s, the main strategy was to place environmental issues onto the political map and create a distinct environmental policy domain. The approach shifted with the 1988 Environmental Policy for the 1990s bill, however, which articulated clearly the need for more integrated modes of environmental policymaking, emphasizing preventive and cross-sectoral approaches (Kronsell, 1997). The importance of EPI in sectoral policymaking was continuously stressed in several ensuing government bills. However, it was not given a strong impetus and political weight until Mr Göran Persson, elected Prime Minister in 1996, declared that ecological modernization was to become a new profile issue for the Social Democratic government. Persson declared to the parliament that 'Sweden should be a driving force and a model when it comes to efforts to achieve ecological sustainability'. The tradition of building a welfare state, popularly called the 'people's home' (*folkhemmet*), was to be transformed into building a 'green people's home' (Nilsson and Persson, 2003; Lundqvist, 2004).

Following this, a number of EPI measures were implemented in relatively rapid succession (Table 1.1). In 1996 an advisory council to the government suggested a set of measures to ensure better environmental integration (Miljövårdsberedningen, 1996). Following this, a small group of ministers formed a Commission of Ecologically Sustainable Development and produced a range of proposals for EPI initiatives and schemes in 1997. In their report 'Sustainable Sweden' they called for an investment programme for a transformation of infrastructure (the 'Ecocycle Billion'), another one for an 'ecocycle transformation' of the energy system and a third 'local investment programme' to push renewable energy and other environmental initiatives in municipalities (Delegationen för Ekologiskt Hållbar Utveckling, 1997).

Regarding legislation, a new Environmental Code was already under development. Other measures included setting up fifteen 'national environmental quality objectives' (NEQOs) as a framework for Swedish environmental policy (Prop, 1997/98:145), implementing environmental management systems (EMSs) for all public agencies systems, accompanied by monitoring and reporting schemes and annual progress reports on ecologically sustainable development delivered to parliament (Eckerberg, 2000). Also at the regional and local levels a range of EPI measures were introduced, such as environmental integration in regional development policies and impact assessment of comprehensive municipal planning. In the early 2000s the government took some initiatives that more explicitly dealt with the broader concept of sustainable development. In 2002 the first national sustainable development strategy was presented (revised in 2004 and 2006) (Skr, 2003/04:129). In 2004 a coordination unit for sustainable development was established in the Prime Minister's Office. And in 2005 the Ministry for the Environment was expanded to include energy, housing and planning issues and renamed the Ministry for Sustainable Development (although in Swedish the

Table 1.1 *Milestones in Swedish EPI*

1988	Ministry for the Environment is created
1988	The Environment Bill establishes principles of 'sector responsibility'
1994	Government Communication declares that all policy should be subject to environmental assessment
1996	Prime Minister presents vision of 'green welfare state' (*det gröna folkhemmet*)
1996	Environmental Advisory Council report on EPI, leading to a Commission of Ecologically Sustainable Development of ministers that carry forward suggestions for measures
1996	Legal generic sector responsibility established for central government authorities
1997	Environmental management systems (EMSs) adopted in government
1997	The first annual sustainable development report from the government to parliament
1998	Local investment programmes for environment and employment
1998	Special sector responsibility for 24 government authorities
1999	The Environmental Code enters into force
1999	Fifteen national environmental quality objectives (NEQOs) introduced (a 16th added in 2004)
2001	Green public procurement for government authorities and public bodies (*EKU-verktyget*) is launched
2002	First version of a National Sustainable Development Strategy), revised in 2004 and 2006
2004	Sustainable development coordination unit established in Prime Minister's Office
2005	Ministry of Sustainable Development established

new name, *Miljö- och samhällsbyggnadsdepartementet*, has a somewhat different connotation). This ministry set-up, however, was short-lived. In 2006 the Social Democrats lost the elections and the new Liberal–Right coalition government shifted back to a conventional 'Ministry for the Environment'.

Three linked measures for EPI are particularly interesting in the context of national policymaking: 'sector responsibility', NEQOs and EMSs for governmental organizations. The principle of sector responsibility was first agreed upon in 1988 and was given formal status in 1995 by being included in the legislation regulating the government authorities. Linked to the sector responsibility was the establishment of the 15 NEQOs through which the principle became substantiated with identification of environmental issues and concrete targets. The cabinet identified 24 government agencies with a *specific* responsibility for integrating environmental concerns in their activities as well as promoting ecological sustainability in their respective sectors. Within the framework of the NEQOs, these agencies were to propose environmental objectives and assess the environmental aspects of the sectoral activities. Follow-up, monitoring and evaluation are main instruments in the implementation of the NEQOs, through so-called 'control stations', checkpoints in time to account for progress. Furthermore, the results of monitoring are shared with the various sector actors. In addition, all agencies

Table 1.2 *The Swedish national environmental quality objectives*

1	Reduced climate impact
2	Clean air
3	Natural acidification only
4	A non-toxic environment
5	A protective ozone layer
6	A safe radiation environment
7	Zero eutrophication
8	Flourishing lakes and streams
9	Good-quality groundwater
10	A balanced marine environment, flourishing coastal areas and archipelagos
11	Thriving wetlands
12	Sustainable forests
13	A varied agricultural landscape
14	A magnificent mountain landscape
15	A good built environment
16	A rich diversity of plant and animal life

Source: www.miljomal.nu

were given a *general* responsibility to contribute towards the achievement of the NEQOs and some were given responsibility for a specific objective (the Swedish Environmental Protection Agency (SEPA) is responsible for 9 out of 15 objectives) and/or a cross-cutting aspect such as cultural environment, public health or biological diversity (Miljömålsrådet, 2004). In the updated policy bill from 2005, the government recognized a need to further enhance the connections between the NEQOs and the specific sector responsibility, and reduced the number of agencies with specific responsibilities from 24 to 18 (Prop, 2004/05:150). However, the two systems remain separated. A limited number of agencies are assigned as 'Environmental Objectives Agencies', each being responsible for a particular objective. In this new system, SEPA is in charge of 10 out of the 16 NEQOs (Table 1.2) and six other agencies are responsible for each of the remaining ones. The specific sector responsibility is then assigned to 18 agencies and covers all relevant NEQOs across these agencies. If readers of this book find this all quite messy, it may come as some comfort that government officials do too!

More or less in parallel to the development of the NEQO system and sector responsibility, since 1997 the government has committed itself to implementation of an EMS. In the first round, all major governmental agencies implemented this system. According to a recent internal evaluation, however, the effectiveness has been limited, in particular because the EMS has been primarily concerned with the direct effects of daily office activities, whereas the more important indirect effects from the agencies' core policy implementation were left outside the system. In 2001 the government also decided that Government Offices (the ministry level) would implement such a system, and that they should pay more attention to indirect effects through impact assessment procedures:

Concern for the environment and health shall be natural parts of the Government Offices' activities. The Government Offices shall pay attention to and examine possible consequences for the environment in the development of proposals for government decisions. (Miljödepartementet, 2004, pp3–4)

Hence the Swedish governmental system has witnessed a considerable accumulation of institutional measures with the intent of promoting EPI at the national policy level. This study is, however, also concerned with understanding the implications of the mainstream set-up of the governmental system and the policy procedure. Our contention is that the ingrained institutions and processes are likely to be far more influential on decision making than newly imposed measures.

Features of the Swedish Governmental System

To understand what might have influenced Sweden's response to EPI, our study adopts an institutional approach and focuses in particular on institutions that influence sectoral policymaking in the central governmental system. Swedish public administration is often seen as a well-oiled machine and an integrated organization. Sweden is a relatively homogeneous society, with an informed public debate and well-organized interest groups. Sweden developed a skilled bureaucracy early in history which, compared to most other countries, is often considered quite competent and efficient (Pierre, 1993). Like the other Nordic countries, Sweden is known for its strong welfare state. Private business – but also labour unions and other interest groups – have influenced public policy formulation ever since the early days of the Social Democrat governments (1932–1976, 1982–1991 and 1994–2006).

What does the policymaking system look like in practice? The government sets policy objectives and allocates the budget. Cabinet ministers are collectively responsible for the government's decisions, and the ministries function as coordinating and planning secretariats and are relatively small by international standards. The ministries are responsible for policy formulation, drafting bills and representing the Swedish government in EU working groups. At the ministry level, an important coordination mechanism is the 'joint drafting' procedure, stipulating that all policies that have a potential impact on other ministries' competencies should be developed in consultation with these ministries. Under the ministries, agencies implement policy decisions and also monitor, analyse and report to the government on the development in their sectors. The agencies, considerably better staffed than ministries, are by tradition and constitution independent of their respective ministries in their day-to-day activities, but they are generally consulted in the policymaking process. 'Ministerial ruling' – when the ministers interfere with implementation – is not allowed under the Swedish constitution.

At regional and local levels, two bodies should be mentioned. The county administrative boards are the regional bodies that implement national policy. The municipalities, in contrast, hold separate elections and have considerable autonomy in legal matters and in the management of financial resources. The municipalities became a primary force in society when the welfare state expanded during the 1960s and 1970s (Gustafsson, 1998). However, regional and local levels of government are not direct objects of this study.

The government deploys 'committees of inquiry' to comprehensively review the state-of-the-art knowledge of the policy issues at hand and make proposals to the government. These committees can be composed of parliamentarians and/or agency officials and experts and often bring in actors from outside the government, such as industrial organizations, academia, labour unions and environmental NGOs. Committees also rely on agencies as well as external expertise to provide assessments and research to the committee. Committees report to the ministries, which then draft the bills, after consultation and negotiation between all ministries that are affected by the policy (the joint-drafting rule). The government then presents the draft bill to parliament, which discusses it in its standing committees or working groups. By this time, the government should have been able to sufficiently anchor it politically to get a parliamentary majority to vote in favour. The exact sequence, however, varies depending on the type of committee. In 'parliamentary' committees, political negotiations between the parties are the basis for the recommendations given to the government, implying that the proposal is already backed by a parliamentary majority. In 'expert' committees, the political negotiation takes place after the ministry has received the proposals and prepared a bill.

A special feature of Swedish policymaking is its consultative policy style (Richardson, 1982). In addition to the above committees, this style includes the 'remiss' system, in which various public agencies, societal organizations and interest groups are consulted before a government bill is presented in parliament. Moreover, a rule of public access to official documents facilitates the transfer of information from government bodies to those affected. Hence there is a tradition of public openness to inputs from various experts and interests as a method for reaching a common understanding of policy problems and their solutions. Environmental groups are consulted through this mechanism and also participate in various informal meetings to formulate new policies. In general, the channels between ministries and agencies and various interest groups are well established. The EU membership since 1995 has, however, posed a challenge to the 'remiss' system as decisions are now taken through partly new procedures and with higher speed than before (Petersson, 1994).

In general, some of these structural and formal relations suggest that Sweden would have relatively good conditions for integrated policymaking. Particularly important institutions to look at are the committees of inquiry and the widespread use of consultative processes. In addition, there is a high degree of political consensus across the political parties – including those on the centre-right – regarding the prioritization of the environment and sustainable development on the agenda. The small Green Party has become influential in recent years due to its balancing role in support of the Social Democratic government. Thus in addition to a strong potential for policy coordination generally, the prominence of environmental concerns would appear to be right up there.

Still, it is an open question how these conditions play out in reality, and what effect they have on EPI. For instance, the committee model has proven to be an arena for learning where policy-relevant knowledge and processes for political decision making can meet (Lundqvist, 1997). However, factors such as the capacities and resources within the committee, its perceived relevance and validity, and the capacity in the policymaking system to make use of the results may influence its potential for EPI (OECD, 2002). The implications of this and the other institutional features of Swedish government for EPI will be addressed empirically in Chapter 6.

Examining EPI in Practice in Energy and Agriculture

Does the Swedish government really provide the actively coordinated, holistic organization that can take into account broader aspects of sustainable development in its policy formation processes? Are the implemented EPI measures (that give Sweden such a good ranking internationally) really reshaping sectoral policymaking? This book will examine these questions empirically by analysing EPI in practice in two sectors, agriculture and energy. These two sectors are interesting for several reasons.

First, they are intimately linked to many of the key sustainability challenges and constitute the major culprits behind some of our most severe environmental problems in Europe. Hence solutions to integration problems are urgently needed. The sectors cause a range of environmental impacts, and face plenty of goal conflicts and difficult trade-offs in dealing with them.

Second, these sectors have been the main targets of more active integration efforts in recent years, in particular through the Cardiff Process. The development of the energy sector was in fact a major driver behind the emergence of nature conservation in policymaking in the first half of the 20th century (Lundgren, 1995). Many see its recent development as a success story, and energy policy is to a large extent presented as an environmental policy. The Swedish energy transition programme from 1997 is an example of how sustainability concerns are at the forefront of the sectoral policy agenda. The agricultural sector is primarily governed by the Common Agricultural Policy (CAP) of the EU, in which the implementation of the policy is taking place through national programmes. Sweden's programme in 2000–2006 is called the 'Environmental and Rural Development Plan' and was predominantly geared towards promoting environmental values. By and large, these sectors would appear to be particular success stories within the broader success story of Sweden in general. In this book, we try to take a critical view of this situation to see whether the reality of EPI (in terms of mainstream sectoral policymaking) match the rhetoric and grand intentions.

Third, these sectors allow for a comparative study as they represent very different issues, institutions, actors and policy contexts. For instance, the agricultural sector is highly regulated but with predominantly small-scale private producers, and the energy sector is deregulated but with mostly large-scale private and public producers. Agricultural policy has been in principle controlled by the EU whereas energy policy has been a national domain. Agricultural policy has been predominantly treated as an administrative concern whereas energy policy has been very high on the political agenda over recent decades. The two sectors also display very different types of environmental problems: the energy sector is responsible for a major part of the air pollution and greenhouse gas problems in Europe, normally through point sources, whereas the agricultural sector is the major source of soil and water degradation and pollution, normally from non-point sources.

The two sectors intersect in the field of bioenergy, which has its own set of environmental challenges, including landscape impacts from its production as well as local and regional air pollution problems from combustion. In addition to the energy and agricultural cases, this book also presents a case on bioenergy policy, which cuts across these sectors. Bioenergy as a concept includes all fuels produced from biological material (biomass) (SJV, 2006) but the focus here is especially on agricultural biomass production. While energy and agriculture are two distinct and established policy sectors, bioenergy cannot be regarded a sector in its own right. Rather, bioenergy – like most environmental issues – is a multisectoral issue, depending as much on energy policy decisions as on agricultural policy decisions. Since studies of EPI are usually conducted on a sector-by-sector basis, the multisectoral nature of bioenergy policy provides a new and important aspect of EPI where in particular inter-sectoral policy coordination is of importance.

A brief introduction to energy policy in Sweden

Sweden is relatively well endowed with energy resources. Electricity production is primarily based on hydropower and nuclear power: 11 nuclear reactors cover about 45 per cent of total electricity use and hydropower delivers about the same. This means that in recent decades electricity production has been almost free of carbon dioxide (CO_2) and also relatively cheap, which has offered a comparative advantage for Swedish industry, enabling it to develop and specialize in electricity-intensive sectors (Johansson, 2006; Nilsson, 2006). Sweden is therefore comparatively electricity intensive (Figure 1.1). With wide areas of forested land, supply opportunities for renewable heat and power production are also comparatively good. Still, energy supply has been a critical issue in Swedish contemporary political history and has over the past 40 years been coloured by political lock-ins and conflict. In the 1950s and 1960s conflicts over hydropower expansion was the major issue, although protests from conservationists began as early as the 1910s (Vedung and Brandel, 2001). And since the early 1970s the question of nuclear phase-out has been a central point of conflict in Swedish politics (Anshelm, 2000).

Energy policy has three overarching goals: security of supply, industrial competitiveness through low prices and environmental sustainability. Under these, the government pursues several objectives: increasing the share of renewable energy, phasing out nuclear power, developing the functioning of the electricity market, enhancing the grid capacity to promote cross-border electricity trade, and reducing carbon dioxide emissions. The majority of Swedish electricity production used to reside within the governmental agency Statens Vattenfallsverk, which was made into a fully government-owned company, Vattenfall AB, in 1992. It still supplies about 50 per cent of Swedish electricity consumption, but is now operating very much as a major international corporation across Europe. The two other major firms are Eon and Fortum. With the exception of Vattenfall, electricity companies today are often private and owned by foreign interests. The Nordic electricity

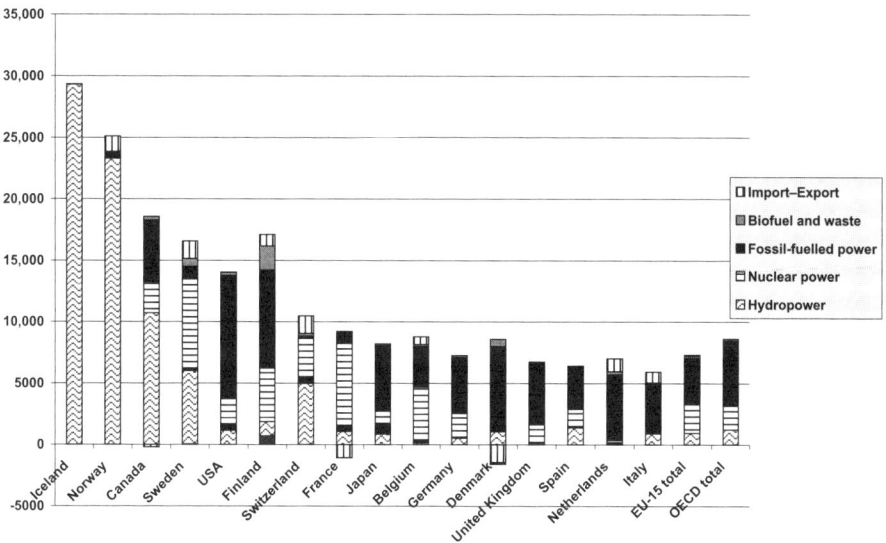

Source: STEM (2005)

Figure 1.1 *Specific electricity production by power source, 2003, kWh/capita*

grids are connected with each other and with Germany, Poland and Russia, and prices are set on a Nordic electricity exchange market, 'NordPool'.

As a result of these changes, today the implementation of energy policy is affected by decision making by many different actors. Interests of producers, distributors, traders and consumers must be taken into account (Nilsson, 2005). Therefore the influence of government policy on the sector is a matter of discussion. The strong price incentive provided through taxes suggests that energy policy in Sweden is still powerful, and a crucial determinant on investments in the sector. Changes in tax incentives or long-term expectations on policy determine not only marginal uses but also long-term investment in, for example, natural gas or biomass energy systems. For instance, differential taxation has facilitated a fuel shift towards renewable fuels in the district heating system, and the use of fossil fuels has decreased significantly in the various sectors (excluding transport – see Figures 1.2 and 1.3). On the other hand, public actors now have a very limited direct role in investment decisions.

The EU has no formal competency over energy policy. However, through adjacent policy areas, we can still witness a profound impact of the European policy context on Swedish energy policy. This involves, in particular, environmental policies such as emissions trading of carbon dioxide and through internal market policies, such as the deregulation of electricity and gas markets (European Parliament and Council of the European Union, 2003 and 1996).

Several of the NEQOs apply to energy policy. The most prominent is arguably 'reduced climate impact' but other relevant ones are 'clean air', 'a safe radiation environment' and 'a rich diversity of plant and animal life'. The Swedish Energy

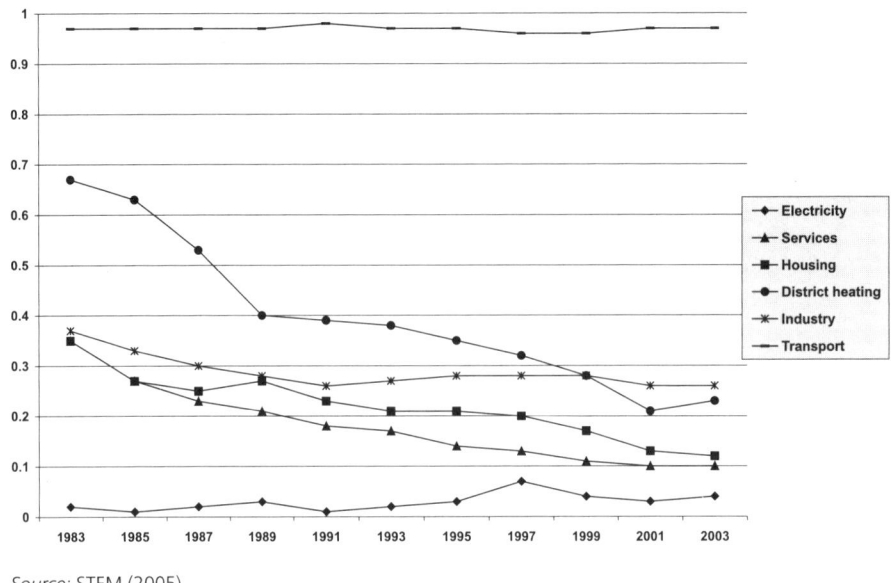

Source: STEM (2005)

Figure 1.2 *Fossil fuels as share of total energy use in different sectors, 1983–2003*

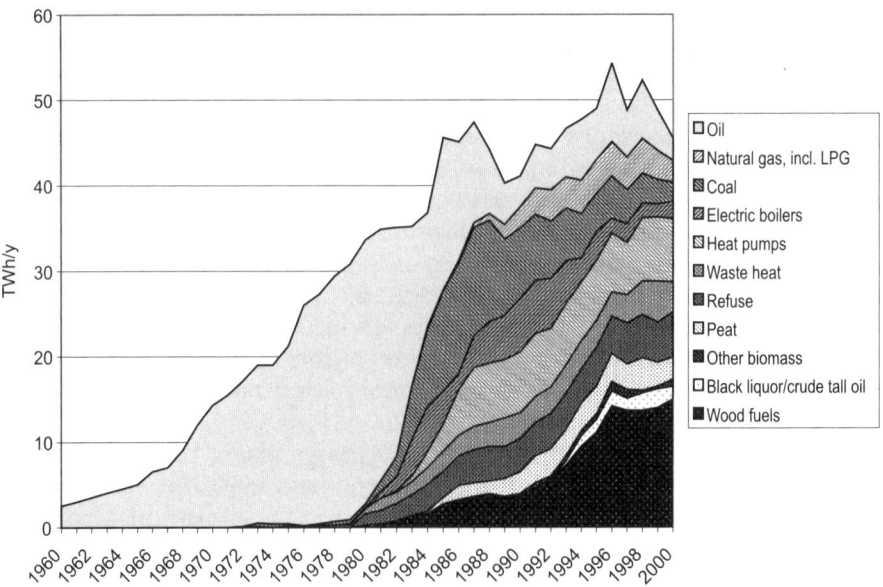

Source: Nilsson et al (2004)

Figure 1.3 *Fuel mix in Swedish district heating, 1960–2000, TWh/year*

Agency (STEM), is not an 'environmental objectives agency', but has a specific sector responsibility for contributing to the relevant NEQOs.

A brief introduction to agricultural policy in Sweden

Swedish agriculture covers 7 per cent of the total land area and 2 per cent of its working population (about equally in farming and the food industry). Sweden's development from an agricultural country to an industrial and post-industrial service society is quite a recent phenomenon. Because of the strong historical links, farmers could until recently expect understanding from the large majority of people living in urban areas. However, even though the idealized picture of farming still exists, the attitude towards farming among the general public is becoming more sceptical, both due to debates on issues such as animal welfare and due to the system of subsidies under the European CAP. The more recent view on farming as simply one economic activity of many has implications for how different environmental initiatives in relation to agriculture are received.

During the post-war period, Swedish agricultural policy became characterized by detailed regulation and large government subsidies aimed at increasing the size of farms, mechanization and measures towards intensification of land use. However, a breakthrough for agri-environmental policy occurred in 1990 with a new food policy including deregulating the internal market for agricultural products (Prop, 1989/90:146). At this time, all subsidies were abolished except those that favoured environmental protection. Protection of biological diversity and traditional farming practices became prioritized issues. The early 1990s can thus be characterized as a new era for Swedish agri-environmental and forest-environmental policies. However, these again took a sharp turn with the EU membership in 1995, which entailed reintroducing large subsidies to agriculture along with detailed regulation (Eckerberg et al, 1994).

Agriculture is hence a sector in which interdependence between Europe and national policymaking is particularly obvious. The CAP (established in 1960) introduced regulation of the internal market (through import fees and structural support) with the objective of securing increased food productivity, reasonable livelihoods for farmers, food security and reasonable consumer prices. However, growing expenses for maintaining the CAP motivated the 'MacSharry' reform in 1992, which included decreased price support for grain, compensation for fallow, decreased intervention prices for beef and increased support for livestock producers based on the number of animals per hectare. It also involved complementary measures such as support for environmentally friendly production methods, wood plantation and a programme for early retirement. In total, there was a shift in focus from price support to direct income support. In 1997 the prospect of a number of new member states from Central and Eastern Europe spurred another reform, the 'Agenda 2000', involving increased market orientation, strengthening competition, food security and food quality, stable incomes for farmers, integration of environmental concern in agriculture, rural development, and overall simplifications and decentralization.

Ecological farming has increased rapidly from the early 1990s onwards. The government had set a goal in 1994 to have 10 per cent of the farming area ecologically managed by 2000 (Prop, 1993/94:157). This goal was also supported

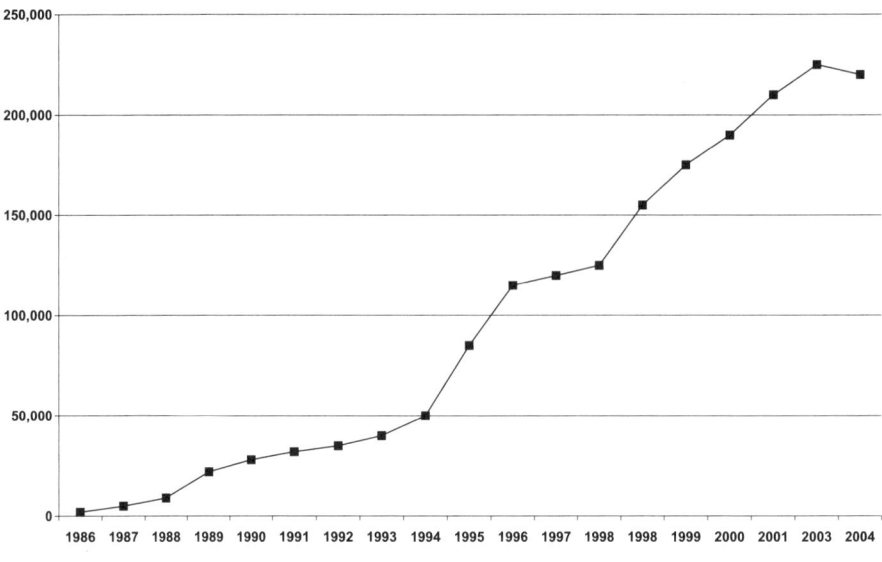

Source: KRAV (2004)

Figure 1.4 *KRAV-certified arable land, 1986–2004, ha*

by the Federation of Swedish Farmers (LRF). In practice, the area cultivated by ecological farming grew rapidly until 2002, but then came to a standstill at about 7 per cent of the agricultural area (Figure 1.4). The government recently proposed a 20 per cent goal to be reached by 2010, accompanied by support measures for certification of farmers and measures to increase consumers' interests in ecological food products (Skr, 2005/06:88).

Several of the NEQOs apply to agriculture policy, including 'high-quality ground water', 'flourishing lakes and streams', 'thriving wetlands', 'zero eutrophication', 'a non-toxic environment' and 'a varied agricultural landscape'. The Swedish Board of Agriculture, under the Ministry of Agriculture, Food and Fisheries (MoA), has sector responsibility for the objective 'a varied agricultural landscape' and also for implementing the Environmental and Rural Development Plan. The Swedish Board of Agriculture, the Swedish Environment Protection Agency and the National Heritage Board are, together, responsible for evaluating the environmental effects of the CAP.

A brief introduction to bioenergy policy in Sweden

Many environmental policy problems – and their solutions – fall between two or more traditional policy sectors, requiring policy coordination. Bioenergy policy is only one example of such multisectoral policy areas. The realization of bioenergy policy goals requires the involvement of both the energy and agricultural sectors to move towards sustainable solutions. Bioenergy offers a potential solution to several of the 'hotspots' of the energy and agricultural sectors. It provides a possibility for the energy sector to decrease the use of fossil fuels and uranium, which in turn would decrease the sector's emissions of CO_2 and its use of non-renewable resources. Energy crops on agricultural land also offer an opportunity for the agricultural sector to become more self-sufficient regarding energy supply (substituting the use of nuclear power and diesel by bioenergy and biofuels) at the same time as it reduces the sector's relatively large contribution to Sweden's greenhouse gas emissions. Swedish bioenergy also concerns the forestry sector because a) there is considerable use and potential to use forestry residues as an energy source and b) cultivation on agricultural land is to a large extent concerned with 'energy-forest crops' such as *Salix* (willow), rather than agricultural crops. However, this section is primarily concerned with the intersections between agricultural and energy policy.

After almost 30 years of developing a Swedish bioenergy policy, bioenergy production has increased, both in absolute numbers as well as its share of total Swedish energy supply. Figure 1.5 shows the expansion of bioenergy in the energy supply.

Agricultural biomass supplies a mere 0.5 per cent of the total bioenergy production. In fact, the lion's share of bioenergy today comes from forestry, by-products from the pulp and paper industry, waste and peat (cf. SJV, 2006). Figure 1.6 shows how *Salix* production expanded in the early 1990s but then went down again. It covers less than 1 per cent of the agricultural land in Sweden, illustrating the lack of policy coordination between the energy and agricultural sector, which will be explored in Chapters 5 and 6.

Source: STEM (2005)

Figure 1.5 *Share of bioenergy (incl. peat) in total energy supply, 1970–2004, %*

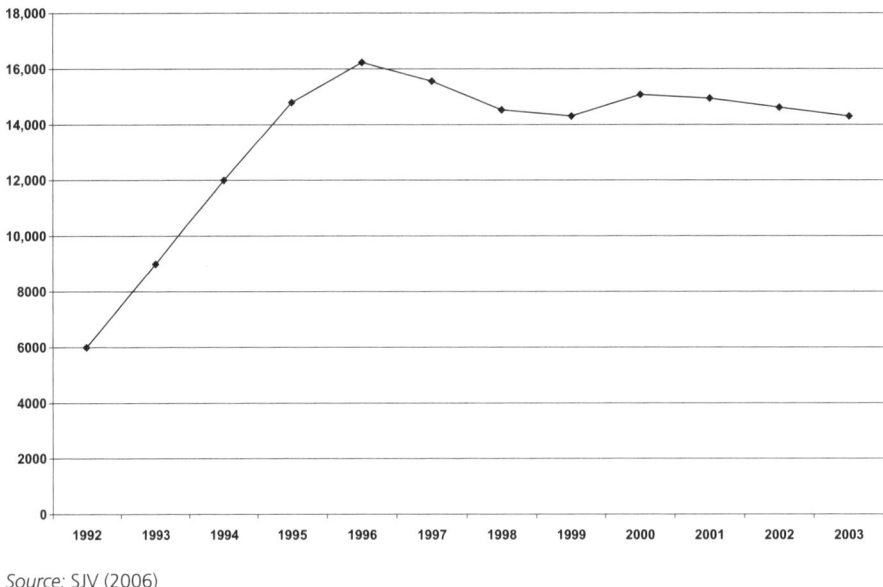

Source: SJV (2006)

Figure 1.6 *Area of* Salix *cultivation, 1992–2005, ha*

Outline of the Book and Research Questions

The book consists of three main parts. The first part is concerned with the conceptual and methodological underpinnings of the book. The practice- or policy-oriented reader might choose to skim through these quickly. The second part gives an empirical account of the EPI patterns in the two sectors in Sweden. The third part provides a synthesis and draws out broader lessons from the cases, as well as a set of policy implications that should be of interest to the practitioner and academic reader alike.

The first, conceptual part consists of Chapters 1, 2 and 3. Key questions here are:

- What are the policy background and the conceptual underpinnings of the EPI principle? (Chapters 1 and 2);
- What different interpretations of EPI exist? (Chapter 2); and
- What would a framework look like for analysing EPI empirically? (Chapter 3).

A starting point for the work is a literature review and a theoretical discussion of the EPI principle and concept. Chapter 2 gives an overview of the relatively rich, primarily European literature on policy integration that has emerged in recent years, and the major existing perspectives with their strengths and weaknesses. It further outlines some of the analytical challenges that have emerged in EPI research and practice to date. This leads up to the main features we should incorporate into an analytical framework for EPI. Chapter 3 is devoted to our analytical framework for understanding EPI and the processes around it. The framework includes the concepts of sector environmental analysis, policy framing and learning, the role of institutions, and what important background variables must be accounted for. We describe the main methods and data needed for the analysis.

The second, empirical part consists of Chapters 4, 5 and 6. Key questions here are:

*   What are the current sectoral environmental outcomes? (Chapter 4);
*   How has EPI in the sectors evolved and how far has this contributed to a shift in policy? (Chapter 5); and
*   What key factors have affected EPI, and, in particular, how is EPI associated with institutional rules, procedures and structures? (Chapters 5 and 6).

Examining energy and agricultural policies and processes in Sweden, including patterns across time, gives a diagnosis of to what extent there really is policy integration. Chapter 4 gives a description of the environmental outcomes of the two empirical sectors (energy and agriculture) based on a comprehensive analysis of the sectors' activities and their implications. It discusses what the most critical environmental aspects in each sector are and characterizes the political problems of EPI in terms of interests and conflicts in relation to environmental issues in the sectors. Chapter 5 describes the evolution of policy framing in the energy and agricultural sectors and to what extent environmental issues have become integrated into sectoral framing. Chapter 6 analyses the institutional landscape and mechanisms at play in the two sectors and provides an analysis of how these institutions, both traditional and new, have contributed to or constrained EPI progress over time.

The last, synthesis part consists of Chapters 7 and 8. Key questions of this third part are:

*   What needs to be done to strengthen EPI? (Chapters 7 and 8);
*   How can institutions be shaped to contribute more effectively to EPI? (Chapters 7 and 8); and
*   To what extent are the Swedish lessons transferable to other countries and levels of governance? (Chapter 8).

Chapter 7 synthesizes the major findings and lessons learned from our analyses and how they relate to existing understandings and policy practice in terms of promoting EPI. Based on the in-depth empirical study, the book concludes by making suggestions about how to enhance the potential for policy integration. In essence: What works and why? Here we will also discuss the extent to which, and in what ways, the Swedish experience is transferable to other countries and levels of governance. Chapter 7 addresses primarily an academic audience and the more policy-oriented lessons are picked up in Chapter 8. Chapter 8 will briefly highlight key findings for those who want a succinct summary of the most important policy implications of our work.

References

Anshelm, J. (2000) *Mellan frälsning och domedag: Om kärnkraftens politiska idéhistoria i Sverige 1945–1999 [Between Salvation and Armageddon: On Nuclear Political History in Sweden 1945–1999]*, Brutus Östlings Bokförlag, Stockholm

Brewer, G. and Stern, P. (eds) (2005) *Decision Making for the Environment: Social and Behavioral Science Research Priorities*, National Academies Press, Washington, DC

Delegationen för Ekologiskt Hållbar Utveckling (1997) *Ett hållbart Sverige [A Sustainable Sweden]*, Regeringskansliet, Stockholm

EC (1998) *Partnership for Integration: A Strategy for Integrating Environment in EU Policies*, European Commission, Brussels

EC (2001) *A Sustainable Europe for a Better World: A European Union Strategy for Sustainable Development*, COM (2001) 264, European Commission, Brussels

EC (2004) *Integrating Environmental Considerations into Other Policy Areas – A Stocktaking of the Cardiff Process*, COM (2004) 394 final, European Commission, Brussels

Eckerberg, K. (2000) 'Sweden: Progression despite recession', in W. Lafferty and J. Meadowcroft (eds) *Implementing Sustainable Development: Strategies and Initiatives in High Consumption Societies*, Oxford University Press, Oxford, UK

Eckerberg, K., Mydske, P. K., Niemi-Iilahti, A. and Pedersen, K. H. (1994) *Comparing Nordic and Baltic Countries: Environmental Problems and Policies in Agriculture and Forestry*, Nordic Council of Ministers, Copenhagen

EEA (1999) *Europe's Environment: The Second Assessment*, European Environment Agency, Copenhagen

EEA (2005) *Environmental Policy Integration in Europe: State of Play and an Evaluation Framework*, European Environment Agency, Copenhagen

European Parliament and Council of the European Union (1996) Directive 96/92/EC 'Concerning common rules for the internal market in electricity', European Commission, Brussels

European Parliament and Council of the European Union (2003) 'Directive 2003/87/EC of the European Parliament and of the Council of 13 October 2003 establishing a scheme for greenhouse gas emission allowance trading within the Community and amending Council Directive 96/61/EC', European Commission, Brussels

Fergusson, M., Coffey, C., Wilkinson, D. and Baldock, D. (2001) *The Effectiveness of EU Council Integration Strategies and Options for Carrying Forward the 'Cardiff' Process*, Institute for European Environmental Policy, London

Gustafsson, A. (1998) *Kommunalt självstyre [Local Government Autonomy]*, Liber, Stockholm

Hertin, J. and Berkhout, F. (2001) *Ecological Modernisation and EU Environmental Policy Integration*, SPRU Electronic Working Paper Series, paper no 72, University of Sussex, Brighton, UK

Johansson, B. (2006) 'Climate policy instruments and industry – Effects and potential responses in the Swedish context', *Energy Policy*, vol 15, pp2344–2360

Jordan, A. and Schout, A. (2006) *The Coordination of the European Union*, Oxford University Press, Oxford, UK

Kraemer, A. (2001) *Results of the 'Cardiff-Processes' – Assessing the State of Development and Charting the Way Ahead*, Ecologic, Berlin

KRAV (2004) Annual Report, KRAV Ekonomisk Förening, Uppsala, Sweden

Kronsell, A. (1997) 'Sweden: Setting a good example', in M. S. Andersen and D. Liefferink (eds) *European Environmental Policy: The Pioneers*, Manchester University Press, Manchester, UK

Lafferty, W. (2004) 'From environmental protection to sustainable development: The challenge of decoupling through sectoral integration', in W. Lafferty (ed) *Governance for Sustainable Development*, Edward Elgar, Cheltenham, UK

Lafferty, W. and Eckerberg, K. (1998) *From the Earth Summit to Local Agenda 21: Working Towards Sustainable Development*, Earthscan, London

Lafferty, W. and Hovden, E. (2003) 'Environmental policy integration: Towards an analytical framework', *Environmental Politics,* vol 12, pp1–22

Lenschow, A. (2002a) *Environmental Policy Integration: Greening Sectoral Policies in Europe*, Earthscan, London

Lenschow, A. (2002b) 'Greening the European Union: An introduction', in A. Lenschow (ed) *Environmental Policy Integration: Greening Sectoral Policies in Europe*, Earthscan, London
Lenschow, A. and Zito, A. (1998) 'Blurring or shifting of policy frames? Institutionalization of the economic-environmental policy linkage in the European Community', *Governance*, vol 11, pp415–442
Liberatore, A. (1997) 'The integration of sustainable development objectives into EU policy-making: Barriers and prospects', in S. Baker, M. Kousis, D. Richardson and S. Young (eds) *The Politics of Sustainable Development: Theory, Policy and Practice within the European Union*, Routledge, London
Löfstedt, R. (2003) 'Swedish chemical regulation: An overview and analysis', *Risk Analysis*, vol 23, pp411–421
Lundgren, L. J. (1995) 'Sveriges gröna historia [Sweden's green history]', in II. Strandberg (ed) *Människa och miljö: Om ekologi, ekonomi och politik [Man and Environment: On Ecology, Economics, and Politics]*, Tidens Förlag, Stockholm
Lundqvist, L. (1997) 'Sweden', in M. Jänicke and H. Weidner (eds) *National Environmental Policies: A Comparative Study of Capacity Building*, Springer, Berlin
Lundqvist, L. (2004) *Sweden and Ecological Governance: Straddling the Fence*, Manchester University Press, Manchester, UK
Miljödepartementet (2004) *Regeringskansliet och miljön. Bilaga till protokoll från regeringssammanträde 2004-05-06 [Government Offices and the Environment]*, Regeringskansliet, Stockholm
Miljömålsrådet (2004) *Miljömålsrådets utvärdering av Sveriges 15 miljömål [Environmental Objectives Council's Evaluation of Sweden's 15 Environmental Objectives]*, Naturvårdsverket, Stockholm
Miljövårdsberedningen (1996) *Miljöarbete i statliga myndigheter: en vägledning om integrering av miljöhänsyn [Environmental Activities in Governmental Agencies: A Guide to Integration of Environmental Concerns]*, Regeringskansliet, Stockholm
Nilsson, L. J., Johansson, B., Åstrand, K., Ericsson, P., Svenningsson, P., Börjesson, P. and Neij, L. (2004) 'Seeing the wood for the trees: 25 years of renewable energy policy in Sweden', *Energy for Sustainable Development*, vol 11, pp67–81
Nilsson, M. (2005) 'Learning, frames and environmental policy integration: The case of Swedish energy policy', *Environment and Planning C: Government and Policy*, vol 23, pp207–226
Nilsson, M. (2006) 'The role of assessments and institutions for policy learning: A study on Swedish climate and nuclear policy formation', *Policy Sciences*, vol 38, pp225–249
Nilsson, M. and Persson, Å. (2003) 'Framework for analysing environmental policy integration', *Journal of Environmental Policy & Planning*, vol 5, pp333–359
OECD (1996) *Building Policy Coherence: Tools and Tensions*, Public Management Occasional Papers No 12, OECD, Paris
OECD (1997) *Environmental Indicators for Agriculture*, OECD, Paris
OECD (1999) *Indicators for the Integration of Environmental Concerns into Transport Policies*, OECD, Paris
OECD (2002) *Governance for Sustainable Development: Five OECD Case Studies*, OECD, Paris
Petersson, O. (1994) *Swedish Government and Politics*, Publica, Stockholm
Pierre, J. (1993) 'Legitimacy, institutional change, and the politics of public administration in Sweden', *International Political Science Review*, vol 14, pp387–401
Prop (1989/90:146) *Om livsmedelspolitiken [On Food Policy]*, Government Bill, Regeringskansliet, Stockholm
Prop (1993/94:157) *Stöd till ekologisk odling och trädgårdsnäringen m.m. [Support for Ecological Farming and Horticulture etc]*, Government Bill, Regeringskansliet, Stockholm
Prop (1997/98:145) *Svenska miljömål. Miljöpolitik för ett hållbart Sverige [Swedish Environmental Objectives: Environmental Policy for a Sustainable Sweden]*, Government Bill, Regeringskansliet, Stockholm
Prop (2004/05:150) *Svenska miljömål – Ett gemensamt uppdrag [Swedish Environmental Objectives – A Joint Mission]*, Government Bill, Regeringskansliet, Stockholm
Richardson, J. (1982) *Policy Styles in Western Europe*, Unwin Hyman, London
SJV (2006) *Bioenergi – Ny energi för jordbruket [Bioenergy – New Energy for Agriculture]*, Report 2006:1, Jordbruksverket, Jönköping, Sweden

Skou Andersen, M. and Liefferink, D. (1997) *European Environmental Policy: The Pioneers*, Manchester University Press, Manchester, UK

Skr (2003/04:129) *A Swedish Strategy for Sustainable Development*, Government Communication, Regeringskansliet, Stockholm

Skr (2005/06:88) *Ekologisk production och consumption – Mål och inriktning till 2010 [Ecological Production and Consumption – Goal and Direction for 2010]*, Government Communication, Regeringskansliet, Stockholm

STEM (2005) *Energy in Sweden: Facts and figures*, Statens Energimyndighet, Eskilstuna, Sweden

Time Magazine (2006) 'Cleaner air over Scandinavia', 3 April, p41

UNCED (1992) *Agenda 21*, United Nations, New York

Underdal, A. (1980) 'Integrated marine policy: What? Why? How?', *Marine Policy*, vol 4, pp159–169

Vedung, E. (1991) 'The formation of green parties: Environmentalism, state response and political entrepreneurship', in J. A. Hansen (ed) *Environmental Concerns: An Inter-disciplinary Exercise*, Elsevier, London

Vedung, E. and Brandel, M. (2001) *Vattenkraften, staten och de politiska partierna [Hydropower, the State and the Political Parties]*, Nya Doxa, Nora, Sweden

WCED (1987) *Our Common Future*, Report by the World Commission on Environment and Development, Oxford University Press, Oxford

Weale, A. and Williams, A. (1993) 'Between economy and ecology? The single market and the integration of environmental policy', in D. Judge (ed) *A Green Dimension for the European Community: Political Issues and Processes*, Frank Cass, London

Yale Center for Environmental Law and Policy (2006) 'Pilot 2006 environmental performance index', Yale University, New Haven, US

2
Different Perspectives on EPI

Åsa Persson

Environmental policy integration (EPI) is a policy principle with great intuitive appeal, as described in the previous chapter. It offers a progressive and 'common-sense' policy approach in that environmental problems are addressed at their source and 'win–win' opportunities for jointly achieving environmental and sectoral objectives are maximized. The most optimistic EPI supporters view it as key to bridging the gap between traditional environmental protection policy and an ecologically sustainable society. However, if we start unpacking EPI conceptually, a range of analytical ambiguities, competing perspectives and, not least, difficult normative issues emerge. A good way to illustrate the more contentious and complicated aspects is to break down the term EPI into its three components.

First, what are the *environmental* objectives or concerns we want to integrate? There may be agreement that the environment as a whole merits more consideration at sector level in relation to economic and social concerns. However, views are likely to diverge on which environmental problems should be prioritized and how internal environmental goal conflicts should be solved (SEPA, 1999). Who possesses authoritative knowledge and who should make the political judgement on which particular environmental issues we should prioritize?

Second, is it always clear what comprises *policy* – environmental and sectoral – in a particular situation? Policies do not always have clear beginnings and ends, nor is policy practice necessarily a reflection of formal policy outputs. Furthermore, there is a significant difference between policy levels such as strategic policy frameworks (for example a national transport strategy), specific policy instruments (for example agricultural subsidies) and technical policy implementation arrangements (for example guidelines for issuing building permits), with obvious implications for what kind of EPI strategy is appropriate and effective. The EPI situation can thus be rather fuzzy. A related issue is whether EPI is primarily a task for elected politicians or administrative staff in government departments and agencies.

Finally, what does *integration* really mean, and does it entail any qualifications? A visionary interpretation is that an integrated whole represents 'more than the sum of its parts' (Eggenberger and Partidario, 2000). A softer approach may accept comprehensiveness of inputs to a decision-making situation as evidence of integration. However, with a more realistic and stringent interpretation it seems difficult to avoid the issue of relative weighting of parts in a decision – in this case environmental versus sector (economic and social) objectives. 'Win–win' opportunities do exist, but there will also be instances of 'win–lose' EPI situations, or perceptions thereof, nonetheless. A key question, then, is whether the environment should have 'principled priority' or not (Lafferty and Hovden, 2003). Regardless of whether such a formal kind of decision rule for EPI is adopted or not, it is clear that integration is a technical term devoid of sensitivity to the power context in which EPI takes place, where environmental interests have traditionally occupied a marginal position.

Initially unpacking EPI in this way suggests that there are many difficult issues lurking beneath the surface. However, rather than discouraging the pursuit of this policy principle, such difficulties just serve to further emphasize the need to better understand what EPI is and how it can be achieved. The purpose of this chapter is to take stock of the existing literature, by contrasting conflicting and complementary perspectives on EPI. The objective is to highlight key conceptual issues that should inform the development of an analytical framework for studying EPI (presented in the next chapter), as well as the design of EPI strategies and tools. This chapter begins with an overview of definitions of EPI and clarification of when and where we can expect EPI to take place. In the next section, the value judgements pertaining to EPI are discussed, in particular whether EPI has a rational or normative justification. Having clarified the conceptual and definitional issues around EPI, the next section summarizes the main strategies, measures and key factors for achieving EPI, based on how the EPI problem is framed. Finally, the chapter is concluded by identifying key gaps in the current knowledge about EPI, to inform the construction of an analytical framework presented in Chapter 3.

Basic Perspectives on EPI: What, When and Where?

The previous chapter described the background and evolution of EPI as a policy principle and its application in practice. In this chapter, the evolution of EPI at a conceptual and analytical level is summarized. EPI has in recent years been firmly established as an academic research topic, and detailed reviews of its intellectual history are available (Lafferty, 2002; Persson, 2004). From the policymaker side, organizations such as the European Environment Agency (EEA, 2005a) and the OECD (2002a) have shown interest in learning from and synthesizing the existing literature. Generally, these literature reviews and best-practice guides suggest that EPI is really a complex of governance problems that requires multifaceted strategies and tools. As a consequence of this broadness, it has been suggested that EPI is interpreted quite differently by various national governments. For example, Lenschow (2002a) states that in the Netherlands EPI is perceived as a bottom–up attitudinal change, in Germany it is perceived as a chance to correct market failures, and in the UK it is perceived mainly as an impetus for rationalizing government and eliminating contradictions. Due to this diversity, it is important to go back to the basics with this concept.

The purpose of studying EPI

Why are we interested in understanding and analysing EPI? There are at least three different purposes for EPI studies, with different implications for what kind of definition and analysis are required (see also Lafferty and Hovden, 2003, p5):

1   *Does EPI take place in the current state of affairs?* This question is interested in the status quo governance situation and the entry point is the traditional, mainstream sector policymaking process. There are two main subsets of questions. First, with a given set of criteria, has EPI been achieved or not? This kind of study thus requires qualifying EPI and making it measurable in some way. Second, what are the main barriers to EPI in current policymaking? Such a study requires greater understanding of the policy context in which EPI is introduced, such as institutional decision-making routines, power relationships and external pressures.
2   *How can EPI be actively pursued?* Rather than focusing primarily on the policy context or assessing EPI in practice, the purpose can be to develop tools for EPI, ranging from high-level strategies to simple checklists for everyday use. To date, proposed EPI tools have been of a generic nature, by creating opportunities for environmental consideration rather than dictating certain environmental choices.
3   *What would we like EPI to result in?* It is also possible to devise a study for specifying an ideal state of affairs, in other words with a normative purpose. This could encompass both how an environmentally integrated policy process should be conducted and what substantive decisions are needed to achieve a certain sectoral environmental performance.

Of course, these purposes are complementary rather than conflicting. However, they demonstrate the importance of being aware of whether a descriptive or normative perspective is taken, as well as a contextual or a tools-oriented approach.

Defining EPI – A set of tools, a paradigm or a principle?

There are several connotations to EPI. In the policymaker community, it might be seen primarily as a set of concrete tools (for example environmental policy appraisals, environmental management systems or green public procurement) or as a broader agenda for policy change to which the cumulative application of tools contribute. It has also been seen as a transitory means towards realizing an ecological modernization paradigm and achieving a new form of cooperative policymaking with positive-sum solutions (Hertin and Berkhout, 2001). Among the more precise definitions, we find those that define EPI in terms of its aims (Collier, 1994) or criteria that should be met (Lafferty, 2002; Lundqvist, 2004). Since these aims and criteria relate to the issue of priority of environmental versus sector objectives, we will return to them in the next section.

A coherent account of EPI – on both the theoretical and empirical levels – is difficult if it is concurrently seen as a set of concrete, short-term tools and as a broader societal paradigm for long-term change. Therefore it is useful to adopt a definition like Lenschow's (2002a, p6): 'EPI represents a first-order operational principle to implement and institutionalize the idea of sustainable development.' (Other such first-order principles for sustainable development include intergenerational equity, precaution and long-term planning, see, for example, Bomberg, 2004.) Defining EPI as a principle allows us to see how it can – with varying success – be translated into decision-making tools that may or may not contribute to a broader paradigmatic change.

What should be integrated in sector policy?

Having defined EPI as a principle, it is still not clear what exactly should be integrated in sector policy. In general, the EPI literature refers to the terms environmental or ecological *objectives* or *concerns* (if not the overall term 'policy') as the 'objects' to be integrated in sector policy. However, these have not been elaborated upon further. Most likely, they can range from broad and long-term aspirations (for example a sustainable transport system), to quantified and timed targets (for example reducing carbon dioxide emissions by 20 per cent by 2020), and finally to specific requirements (for example protecting a certain wilderness area from road construction projects). Arguably, many environmental objectives will be perceived as both a restriction (for example reducing existing fossil-based energy generation capacity) and an opportunity (for example developing a renewable energy industry) when presented to sector policymakers.

To reduce conceptual complexity, it seems useful to delimit EPI to refer to the integration of environmental objectives, rather than also encompassing integration of policy actors, of institutional levels and of more extensive time and space perspectives, as proposed by Liberatore (1997; see also Steurer, forthcoming). A recent debate in Sweden raised the issue of integrating sustainability objectives rather than just environmental ones, in the sense of adding economic and social objectives. However, this may overcomplicate the issue, as described in Chapter 1. The basic idea of EPI is that it should *lead to* sustainable development (see Lenschow's definition above) by putting environmental objectives on a par with existing economic and social sector objectives. Infusing economic and social objectives too seems superfluous, or potentially defeating the purpose of EPI. Furthermore, assuming that 'sustainability' is a separate (and internally coherent) policy area from sector policy in this way is dubious. Nevertheless, on the same basis as EPI, it can of course be argued that economic and social objectives that have not been given much attention in traditional sector policy should be focused on, for example public health issues, gender equality and disability issues. But it seems conceptually sounder to keep sustainable development as an overarching and ultimate goal rather than as a coherent policy objective to be integrated.

EPI in the stages of the policy process

Having defined EPI and clarified what should be integrated, the next question is when and how integration can take place. Collier (1994, p45) argues that integration can take place at three different stages in the policymaking process: integration of objectives in policy formulation; translation into policy measures; and implementation by government agencies and other actors. This can be illustrated with examples from EPI in the energy sector:

- Policy formulation – for example, the energy sector is given a new fundamental objective, alongside security of supply and competitive energy prices, to minimize environmental externalities. A quantified and time-set target can also be introduced, for example reduction of CO_2 emissions from the energy sector by 20 per cent by 2020;
- Policy measures – for example, a new investment subsidy for wind power, a CO_2 tax or research funding for solar power; and
- Policy implementation – for example, environmental impact assessment in the permitting of energy production plants, green electricity purchasing by government authorities or environmental eligibility criteria for investment support.

The classification of EPI actions into these stages is debatable, but the point is that EPI can take place at more strategic or at more operational stages and policy-making levels. Collier (1994) noted that it is often easier to reach consensus in the policy formulation stage, regarding higher-level integration objectives. Disagreements may not emerge until the implementation stage, where explicit decisions and trade-offs need to be made, with visible cost redistribution and change of actor responsibilities. This may reflect the inherent need for simplification and generalization at the level of strategic decision making, but possibly also an intentional delegation of more uncomfortable decisions down the government hierarchy. In the context of EU transport policy, Hey (2002) found that EPI is relatively strong at the agenda setting stage, which is supranational and strategic in nature and where environmental coalitions are influential. This stage is decoupled from the decision-making stage, however, where vested interests dominate, resulting in weak environmental decisions.

The role of sector context for EPI

Apart from the sector policy stages, the definition and characteristics of the sector have a bearing on the pursuit of EPI. Generally, particularities of the sector context have been less explored in the EPI literature than the organization of policymaking. However, based on its experience with EPI, SEPA (2000 and 2004) has found that how the sector is defined and delimited is important for the division of responsibility for EPI and the setting of sector environmental objectives. They argue that a sector can be defined by 'a group of regularly interacting actors', 'a set of activities with the same implications' or 'a statistically delimited segment of society'. In particular, if a life-cycle approach is taken to measuring sector environmental performance, the match between the sector activities and the way in which sector policymaking is organized in government can be problematic. For example, is energy efficiency in the agricultural sector a matter for energy or agricultural policymakers? This highlights the need for systematic sector environmental assessments as a basis for analysing and implementing EPI, both for understanding major sector environmental impacts and possible internal environmental goal conflicts and for clarifying policy mandates. Thus while it is important to understand EPI from the policymaking end of the spectrum, it is also important to understand the nature of the ultimate target for EPI, namely sector environmental performance. In the context of this book, Chapter 4 presents such a study.

SEPA (2000) has also noted how the characteristics of sectors are significant for the implementation of EPI. Examples include the 'proximity' of the sector activities to environmental events and processes, the existing competence and legal basis for intervening in sector activities, and the technological potential for genuine 'win–win' solutions (see Hertin and Berkhout, 2001). Hey (2002) also proposes that sectoral regulatory capacity is a key factor for EPI. This capacity depends on the financial resources, legal competencies, legitimacy and target group support, and information on the sector regulatory authorities. If capacity is weak, there is a risk that there will be only symbolic or defensive EPI, or a general lack of policy and regulation. Again, the different relationships between sectoral policymakers and (non-state) sector actors emphasize the need for more sensitivity to the particular sector context when working with EPI and the problems with assuming that a uniform EPI approach in all sectors is going to be effective.

The procedural and substantive dimensions of EPI

It is clear that EPI can be studied in the process of making sector policy, in sector policy outputs, and ultimately in sector environmental impacts (see also Persson, 2004; Jacob and Volkery, 2004). The *policy process* perspective on EPI is interested in how policy is made and organized and how opportunities and requirements for environmental consideration can be built into the process, through tools such as environmental policy appraisals, environmental correspondents in sector government departments and setting sector-specific environmental targets. The *policy output* perspective is interested in how sector policy measures encourage or coerce sector actors to change their behaviour in a more environmentally friendly way, through regulations, financial incentives or information. Finally, the *sector impacts* perspective on EPI is focused on examining how changes in behaviour lead to improvements in sector environmental performance.

Naturally, these perspectives on EPI are not mutually exclusive, but rather complementary and interrelated in that the latter would be a consequence of the former. However, they suggest that the focus can be on either *procedure*, creating opportunities and systems for applying the EPI principle in individual decisions, or *substance*, the result of applying the principle in terms of desirable environmental properties of a decision. Lenschow (2002a) suggests that in the EU context, EPI has been understood so far mainly as a procedural principle. This implies that environmental consequences should be recognized and assessed through various formal procedures and the decisions adjusted accordingly. However, the interpretation of EPI as a procedural principle also has some problems in terms of addressing the relationship between integration process and substantive policy output, and eventually environmental impact:

> *In the absence of clearly defined policy goals, indicators and timetables, however, there remains ample room for sectoral policymakers to evade such substantive*

environmental responsibilities. The integration process currently faces the challenge of ensuring that substance follows from procedure. (Lenschow, 2002a, p7)

Thus a key initial question is how effective procedural EPI tools used in the sector policy process are in translating into substantive EPI in the policy outputs. There is a risk in assuming that EPI procedural tools are a proxy for better environmental decisions, since tools may be very ineffective or not used in practice. Furthermore, if a study is measuring and evaluating EPI success, it has to be decided whether it is acceptable that a process with a high degree of EPI can produce a policy output with little positive environmental impact (see also Lafferty and Hovden, 2003, p6). For example, when integrating energy and environmental policy there may be good access to the policy process for environmental actors, good communication processes and a good information basis, but in the end other objectives such as security of supply and industrial competitiveness may be prioritized.

A second key question is how effective the environmentally integrated policy outputs are in changing the behaviour of sector actors and resulting in more positive sector environmental impact. This causal chain can be very complex to study, as it raises the issue of policy effectiveness as such, not just the EPI element. In other words, how influential are policy measures in changing environmental performance, as opposed to factors such as market trends and new technologies? And how do environmental pressures from changed production and consumption behaviour lead to changes in environmental impact, taking into account natural fluctuations in the environment, external shocks and problems of attribution?

In summary, to understand what EPI is, it is useful to break it down conceptually. A basic question is what the purpose of an EPI study is; to understand how it is currently implemented in sector policy, how it can be applied through different tools and/or what environmental changes we want it to result in. Regardless of purpose, however, it was argued that defining EPI as a principle is appropriate, in order to distinguish it from its potential consequences in terms of tools or a broader process of change. There is also a need to clarify what exactly is integrated. To this end it has been found that environmental objectives and concerns is the most logical and practical answer at this point, rather than including dimensions such as actors or time horizons. Furthermore, it is clear that EPI can take place at different stages in policymaking and that it is important to understand how the nature and characteristics of the particular sector in question may influence EPI success. Finally, the procedural and substantive dimensions of EPI have been highlighted, as well as the key question how effective procedural tools are in ensuring substance.

EPI and the Hierarchy of Sector Policy Objectives

Having outlined some of the conceptual and definitional aspects of EPI, it is necessary to examine the normative issues and value-based judgements that EPI involves. In essence, EPI represents certain environmental values and is a normative principle. However, a softer approach is also possible, focusing on the organizational rationale of EPI and its potential to rationalize decision making (see Lundqvist, 2004). Below, the spectrum of views is discussed, followed by the questions of 'how much' EPI is appropriate and the democratic issues involved in who should make such trade-offs.

An organizational or normative rationale?

Chapter 1 described how the EPI principle arose from normative concerns that environmental issues were under-prioritized in sector policymaking. Classical political science and public administration literature, however, demonstrates how policy integration and coordination in general rationalize and increase the efficiency of decision making by preventing unwanted impacts and side effects at the earliest possible stage. Underdal (1980) set three criteria for policy integration: *comprehensiveness* in terms of inclusiveness of space, time, actors and issues in the input stage; *aggregation* in terms of using an overarching criterion to evaluate different policy elements in the processing stage; and *consistency* in terms of all the policy elements being in agreement in the output stage. While recognizing the problems and limitations involved, Underdal nevertheless constructed a rationalist ideal of integration. B. Guy Peters (1998) approaches policy coordination from an organizational perspective and in a more practical way. According to him, it refers to 'ensur[ing] that the various organizations [...] charged with delivering public policy work together and do not produce either redundancy or gaps in services' (p5). Rather than setting criteria for what counts as coordination, Peters refers to the Metcalfe scale in which the first step is 'independent decision making by ministries' and the final (and ninth) step is 'overall governmental strategy'.

Importantly, these accounts do not consider the traditionally low status of environmental objectives and concerns in sector policymaking, which justified the EPI principle in the first place. The implication is that EPI is seen as involving a fundamental revision of the traditional hierarchy of sector policy objectives (Lafferty, 2002). This suggests that *environmental* policy integration is a much more contentious undertaking than other forms of policy integration as described above. How has this challenge been met by EPI theorists?

In some accounts the EPI challenge is still framed in largely positive, 'win–win' terms. For example, Eggenberger and Partidario (2000, p204) argue that 'whenever there are two professionals with different backgrounds looking at the same problem' or 'whenever there are two topics that need to be tackled together, there is integration'. A similar argument is that the integration idea is not particularly novel since all policies represent compromises between different values and multiple objectives (Barrett and Fudge, 1981, in Collier, 1994). Hertin and Berkhout (2003), meanwhile, recognize the power structures that lead to prioritization of economic objectives and argue that EPI should be more than just a 'layering' of environmental demands on top of existing policy processes. However, they still argue that if organizational and perceptual barriers were removed, a new cycle of 'positive-sum solutions' and compatible policy aims is possible.

In Collier's account (1994, p36), a predominantly organizational rationale seems to inform the definition of EPI aims, 'removing contradictions' and 'realizing mutual benefits' between and within policies. However, a third aim is to achieve sustainable development, and in relation to this the need to identify trade-offs between environmental and economic objectives is emphasized. In a similar vein, Lenschow (2002) argues that sustainable development – the 'mother concept' of EPI – may rightly assume compatibility between economic, social and environmental concerns at a global level. But at sector and sub-sector levels trade-offs are likely to appear and there may be real winners and losers of EPI. This leads Lenschow to ask, as the EPI principle becomes increasingly operationalized, will the legitimacy of sustainable development collapse? Liberatore (1997, p119) takes this line of reasoning one step further, by stating that 'the concept of integration assumes a form of reciprocity'. This means that if environmental objectives are given lower weight than economic objectives when making trade-offs for the sector, there will not be integration but 'dilution'. Indirectly, then, the dilution argument sets the criteria of equal weight for decisions to qualify as concordant with the EPI principle.

Lafferty (2002; with Hovden, 2003) proposes the most explicitly norma-tive rationale, also resulting in the most explicitly normative criteria for EPI. The organizational rationale, with EPI as merely 'good policymaking strategy', is rejected. The essential difference is that 'the general environmental or ecological element of sustainable development is the most fundamental – the one without which the concept loses its distinctive meaning' (Lafferty, 2002, p2). Further-more, there are 'numerous very real conflicts of interests with respect to many environmental issues' and 'a primary focus on the search for mutual benefits may [...] draw attention away from the fact that environmental policy often affects certain interests in a negative manner' (p12). Due to the environmental privilege supported in political documents such as Agenda 21 and due to the irreversibility of some damage to life-support systems, environmental objectives thus deserve a 'principled priority' and should become principal or overarching rather than subsidiary. Lafferty thus qualifies EPI by stating that it should imply:

- *the incorporation of environmental objectives into all stages of policymaking in non-environmental policy sectors, with a* specific recognition *of this goal as a guiding principle for the planning and execution of policy; and*
- *an attempt to aggregate presumed environmental consequences into an overall evaluation of policy, and a commitment to minimize contradictions between environmental and sectoral policies* by giving priority to the former over the latter. (Lafferty, 2002, p13, my emphasis)

So, should an organizational or normative rationale guide EPI, and should EPI be qualified with a 'principled priority' criterion? Arguably, this is not an either-or question, but an intermediary position along this spectrum of views is possible. The normative perspective highlights the need to recognize conflicting interests and 'win–lose' situations, the implicit weighting of objectives that integration involves, and the importance of adopting a critical approach to EPI, in which 'dilution' outcomes are also considered. On the other hand, a strict criteria-based EPI approach may not be necessary. Apart from formulation and measurement problems, a strict criteria-based approach may also preclude the recognition of first steps towards EPI.

Interestingly, while both the organizational and normative rationales for EPI assume that it is a desirable principle for policymaking, there have been very few arguments against integration. Is there a case for increased specialization in policymaking as a way of achieving a better environment? Weale (2005, p106) contends that 'it is not obvious, to say the least, that policy integration is the right way to go to secure environmental improvements', for two reasons. First, he argues that the narrow attention spans of policymakers and the difficulties of predicting how a policy may cause environmental change means that environmental policy effectiveness is higher when attention is concentrated on a limited set of problems where technical solutions are known or can be anticipated. Second, an environmental government department could never assert control and power over sector departments because it lacks an essential instrument of control, such as a treasury department's control through budget constraints and the 'simple measuring rod of money'. However, Weale's critique is arguably targeted at how EPI efforts are implemented and performed (to be discussed in the next section), rather than at EPI as an idea necessarily. Furthermore, his empirical examples point to the successes with 'end-of-pipe' environmental regulation and not how to address sector driving forces, such as increased transport demand or urban sprawl. One of the purposes of EPI is to build upon established successes of reactive, 'end-of-pipe' policy and find more proactive solutions, effectively turning problems into opportunities.

The 'optimal' level of EPI

It is thus clear that there is both an organizational and normative rationale for pursuing EPI and making it a legitimate cause. The next question, then, is to what extent and at what cost should it be pursued (provided the cost can somehow be estimated)? What is an appropriate level of EPI, in terms of revising the traditional hierarchy of policy objectives and ensuring that this is possible through providing information, resources for analysis and possibly compensating losers? If a 'principled priority' stance is taken towards EPI, such as that of Lafferty, a certain non-negotiable minimum level is set. It is also a high level, considering today's typical prioritization within sector policymaking, demanding a rather dramatic reallocation of staff and time resources.

Underdal (1980) and Collier (1994), meanwhile, maintain that the principle of Pareto optimality could guide how trade-offs between environmental and economic objectives should be determined. With regards to his *aggregation* criteria, Underdal admits there is a problem with the simplest solution, which is a 'sum-perspective' of weighing costs and benefits of integrating various policy objectives. First, inter-party comparison of utilities is always difficult, especially when environmental values are involved. Second, a 'sum-perspective' could have significant distributional effects. He concludes that aggregation should be conducted in a Pareto optimal way. This restriction would imply that EPI should lead to allocation of resources whereby no one can be made better off without sacrificing the well-being of at least one person. However, both authors recognize that the Pareto criterion is very difficult to apply in practical terms.

Underdal (1980) has also commented on how far the *comprehensiveness of inputs* criterion should be taken. He admits that the optimal scope of inputs to a policy decision is difficult to define, considering that increased inputs (in the case of EPI, environmental analyses and assessments) also involve higher costs. However, a 'rule-of-thumb' is suggested: 'policy comprehensiveness should be measured in relation to the fund of knowledge about policy consequences available at the decision time' (p161). He argues that 'one would expect an inverse relationship between comprehensiveness on the one hand, and aggregation and consistency on the other; other things being equal, the more comprehensive a certain policy, the more centrifugal forces will be at work' (p163). Likewise, perfect policy integration may not be desirable from a cost-effectiveness perspective: 'policy integration should be pursued up to the point where marginal cost of integration effort equals marginal gain from policy improvement, and no further' (p165).

To a large extent, these responses to the 'optimality' questions are purely theoretical. In practice, valuation of costs and benefits of changing sector practices, and also costs of adapting policymaking to EPI, is difficult. Therefore more intuitive solutions to how much sector policy should be 'greened' are required in practice, and the question is likely to be determined politically rather than technically. There is one important limitation to the extent to which EPI should be pursued, however – the norms of democracy.

Who integrates? Democratic aspects of EPI

As described above, Lafferty's definition of EPI represents one of the most explicitly normative views. However, a restriction is also placed on the 'principled priority' of environmental objectives. According to the norms of democracy, environmental objectives cannot automatically override other societal objectives, but must be subject to a democratic decision-making process. Hence Lafferty (2002) concludes that 'the ultimate trade-off for EPI is between existing democratic norms and procedures on the one hand, and the goals and operational necessities of sustainable development on the other' (p15). The democratic aspects of EPI are surprisingly not elucidated by many other authors, except for later publications by Lenschow (2002b and d).

In a system of representative democracy, the key issue is whether EPI-related decisions and judgements should be made by elected politicians or bureaucrats. The Swedish EPI experience has suggested that, in the absence of an overall priority setting of environmental and societal goals by the government, trade-offs are dealt with at the administrative level, sometimes by an individual civil servant (SEPA, 1999 and 2003). This has obvious implications for the accountability for EPI.

From an effectiveness perspective, Peters (1998) also discusses the appropriate division of responsibility for EPI between higher political levels and lower administrative levels. He distinguishes between policy coordination and administrative coordination. The latter represents a bottom–up orientation towards making government policy outputs more coherent and easily absorbed for the target group, and thus stems from an implementation perspective. The former orientation rests on the assumption that it is more effective to coordinate from the very start, in a top–down manner, and that clear overarching priority setting is more important than implementation concerns at the policy formulation stage.

Furthermore, Peters discusses the ownership aspect of policy coordination (see also Schout et al, forthcoming). Is it most effective when it is imposed on administrators or when they are left to bargain among themselves? Different views lead to different conclusions on appropriate means: imposition is most effective in a hierarchical organization while bargaining is the natural option in market and network organizations. Like Underdal, Peters also enters into a discussion of the desirability of increased policy coordination, as it makes it more difficult to hold someone accountable (see also Meijers, 2004). Accountability always needs to be ensured when taking measures to improve coordination.

Intra- and inter-sectoral EPI

Related to the issue of ownership and whether a top–down or bottom–up approach is preferable, two forms of EPI have been identified: intra-sectoral EPI and inter-sectoral EPI (Lundqvist, 2004). Lafferty refers to these as vertical environmental policy integration (VEPI) and horizontal environmental policy integration (HEPI) respectively. VEPI refers to 'the extent to which a particular governmental sector has taken on board and implemented environmental objectives' (Lafferty, 2002, p16). While others use the term vertical for integration between different constitutional levels (see, for example, OECD, 2002b), the vertical axis in this case signifies 'the administrative responsibility "up and down" within the arena of ministerial sectoral responsibility' (p18). According to Lafferty, this type of 'greening' does not need to imply that environmental objectives are given explicit 'principled priority', but that the sector itself develops an understanding of EPI. HEPI, on the other hand, is 'the extent to which a central authority has developed a comprehensive cross-sectoral strategy for EPI' and in this case the issue of principled priority becomes very important. Lafferty proposes that one should understand HEPI as 'insisting on at least equal treatment for the environment vis-à-vis other competing interests' (p18). Lundqvist (2004) proposes a similar interpretation, in that intra-sectoral EPI can take place by simply taking into account environmental objectives in various ways and finding win–win opportunities, while inter-sectoral EPI encompasses a more normative approach where environmental objectives are given equal or more than equal weight in relation to economic and social objectives. In practical terms, implementing HEPI, or inter-sectoral EPI, is associated with the adoption of a national sustainable development strategy or another high-level political document that comprehensively addresses environmental and sector policy. Somewhat paradoxically, HEPI thus seems more in line with the top–down, hierarchical approach to integration (see discussion above), while VEPI is more in line with a bottom–up, network perspective on policymaking.

The value of distinguishing between intra- and inter-sectoral EPI is questionable, however. Because no 'principled priority' of environmental concerns is required in the intra-sectoral form, this suggests that no weighting is needed at the sectoral level and that intra-sectoral integration opportunities are only of a 'win–win' character and do not involve trade-offs. However, such conditions do not seem realistic. Second, it is questionable if it is possible in practice to distinguish between the two dimensions, by tracing a particular policy output to either an internal sector EPI initiative or a cross-sector strategy. Finally, the distinction is based on a rather rigid view of highly separated and self-contained policy sectors.

To sum up, this section has discussed the normative aspects of EPI. It has been explained that when 'win–win' opportunities have been exploited, EPI boils down to a question of weighting environmental and sector objectives. There is a spectrum of views between justifying EPI from a purely organizational and decision-making efficiency basis and justifying it from a strictly normative basis with the implication that environmental objectives should receive 'principled priority'. However, it has been argued that there is no need to choose between these two extremes. Furthermore, different criteria have been explored for the extent to which EPI should be pursued. These are likely to have limited practical relevance, and instead the main limitation is whether EPI is pursued in accordance with democratic norms. A critical issue is whether EPI-related trade-offs are made by politicians or administrators, due to accountability concerns. The question of who should make EPI decisions has also been examined from an effectiveness perspective, with no definitive answer whether a political 'top–down' approach or a administrator-led 'bottom–up' approach is better. There is a parallel to this distinction in that EPI is often classified into either an intra-sectoral or an intersectoral form.

Different Approaches to Achieving EPI

Different interpretations of EPI lead to different understandings of how EPI is best achieved: what the key factors and barriers are and which practical tools are most effective. Four broad groups or approaches can be identified, based on their problem framing and assumptions regarding the nature of policymaking: *procedural*, *organizational*, *normative* or *reframing* approaches. These four approaches are neither mutually exclusive nor exhaustive, but rather show what kinds of variables can be emphasized. Most accounts are well-rounded and portray EPI as a multidimensional challenge, for example the EPI criteria and checklists developed by the OECD (2001a) and EEA (2005a).

The procedural approach

The first steps towards EPI often appear to consist of imposing changed procedures for policymaking or adding specific EPI procedures. Examples include *ex ante* environmental impact assessment, green public procurement requirements, sector environmental reporting requirements and environmental management systems. These could be considered the 'low-hanging fruits' in EPI, but if applied ambitiously there is no reason why they could not result in radical sector policy changes. In any case, advocating these procedural types of tools suggests that the status quo is maintained at the overall level and that the EPI challenge is delegated to sectoral authorities and government departments. Furthermore, these procedural tools are intended to work within the given organizational structure, with given professional expertise and knowledge. Assumptions underlying these tools appear to be a responsiveness to imposed tools among administrators, a view of the policy process as stages in a rational decision-making process, and that the problem with EPI is not *who* should undertake it or *what* should be integrated but *how* it should be.

A procedural tool often identified in the literature is an *ex ante impact assessment* procedure. Strategic environmental assessment (SEA) has been suggested as a promising and practical tool for EPI (Eggenberger and Partidario, 2000; European Eco Forum, 2003; Bina, forthcoming). However, it could be more challenging to apply at a policy level than at a planning level, due to more uncertain and complex activities from which to predict impacts. The EC has recently committed to *ex ante* policy-level impact assessment, whereby economic, social and environmental impacts of policy proposals will be assessed (EC, 2002). Similar initiatives have been taken in several member states (Jacob and Volkery, 2004; Volkery et al, forthcoming). Regarding the effectiveness of impact assessment, simply establishing a procedure may not be sufficient. Rather, it is the interest in, perceived usefulness of and capacity for undertaking environmental assessment that determine its effectiveness in pursuing EPI. These problems of achieving effective assessment in policymaking have been documented for the UK (Russel and Jordan, 2004) and for Sweden (Nilsson, 2006).

Procedures also refer to the institutional *rules of decision making* in a policy system. Rules can refer to the right to set formal agendas, the right to develop policy proposals, and the timing of participation by environmental departments or agencies. As an example, in an examination of EPI in Germany, Müller (2002) found that administrative rules prevent early integration. A formal set of rules for interdepartmental cooperation specifies that the Ministry of Environment can only initiate and negotiate under its own leadership those items that are within its jurisdiction. If the Ministry of Environment wants to cooperate on a policy issue within another ministry's jurisdiction, it has to wait until the competent ministry formally invites them for a ministerial meeting.

To achieve general policy coherence, the *budgetary process* and 'green budgeting' can be an important procedure for promoting EPI in a given organizational structure. It affects all sectors, provides a cyclical opportunity for revision, and operationalizes the government's priorities in a very concrete and often quantitative way (Peters, 1998). In addition, the budget can be restructured to integrate horizontal dimensions, such as environmental concerns (OECD, 1996; UK Cabinet Office, 2000). This can be done through various means, such as including environmental performance objectives and providing incentives, rewards and sanctions for better environmental performance. The budgetary factor also highlights the need for an appropriate allocation of resources and capacity in order for policymakers to carry out EPI effectively (see Weale and Williams, 1993).

Finally, several authors raise the issue of *interaction with non-governmental actors* to achieve EPI, either through formal *consultation and participation processes* or more informal contacts and working relationships. The rationale for increased consultation and participation is twofold: it can make the EPI process more democratic and more efficient by both providing more knowledge and information and increasing the chances for broad acceptance of the policy outputs. The OECD (2002a) states that partnerships between government and business are expected to play a critical role, as well as greater inclusion of the general public. Likely difficulties, however, are lengthy consultation processes and finding appropriate arbitration techniques. Collier (1994) concludes that an open government and an equal level of access to policymakers for environmental lobby groups as for industrial and other lobby groups are conducive factors for EPI.

In addition to problems with effectiveness of procedural tools, another point of critique is that no guidance is given regarding how to deal with the substance of trade-offs between environmental and economic objectives. In other words, these tools can provide the infrastructure for EPI but not neccssarily ensure the normative and substantive decisions required for it (cf. Lenschow, 2002a).

The organizational approach

In addition to, or instead of, imposing procedural tools, an organizational approach can be chosen. This can focus on restructuring the organization of policymaking to address lack of competence and mandate, problems of communication, lack of resources and capacity, and power imbalances. Three underlying assumptions of this approach are that the policy process is a process of communication and bargaining among actors, that organizational identity matters, and that policymaking takes place in a certain power structure.

Virtually all authors concerned with EPI identify a fundamental organizational problem as *sectoral compartmentalization* within government (see, for example, Weale and Williams, 1993; Collier, 1994; Hertin and Berkhout, 2001; Jordan, 2002). Institutional fragmentation (or 'sectoral specialization', 'departmental pluralism', 'functional differentiation') of environmental and sectoral policymaking is the consequence of an efficiency objective: bureaucratic specialization can reduce complexity and increase cost-effectiveness (Hertin and Berkhout, 2001). The negative implication in terms of EPI is that a turf mentality has developed within government – the 'environment' versus the 'sector'. According to Jordan (2002), sectoral compartmentalization has given rise to a tendency towards competition between sector departments to realize their interests. This has become an entrenched philosophy that precedes any 'rational' assessment of a new policy problem. To aggravate the problem, the environmental portfolio has traditionally had a low status (Weale and Williams, 1993). Lafferty (2002, p22) states that 'a separate sectoral environmental authority will rarely, if ever, have the authority necessary to insert environmental objectives into the decision-making premises of other sectoral authorities'. Doern (1993) has examined the potential for the Canadian Department of Environment to become a more central agency. He argues that central space is limited and 'new aspiring entrants [...] must confront and understand the bases of power of those who are already there' (p175). The power resources the department needs to assemble to increase its status are an extended statutory mandate, extended capacity to deal with the increasing volume of decisions, structured contact with other departments, support by external actors in the green policy community, and convincing analytical and scientific capacity for sustainable development. Different responses to this and other organizational impediments to EPI have been suggested, for example adaptation of the organizational arrangements, improvement of coordination and communication processes, provision of incentives through budgeting, and extended interactions with external actors.

Responses to the compartmentalization challenge, in terms of changes to the *organizational arrangements*, can take the form of a) integrating departments and functions, b) establishing new institutions, or c) assigning existing institutions a new mandate, responsibility and accountability. With regards to the first two forms, long lists of possible reforms are provided by Peters (1998) and OECD (1996). For example, committees or boards to monitor sector departments can be established, a minister with a coordinative portfolio can be installed, and 'super-ministries' can be created. However, many regard such institutional reforms with scepticism. Institutional reforms are difficult unless there is a perception of a serious crisis. They are disruptive and it takes time to adjust to the new structure (Lenschow, 2002b; OECD, 1996). In relation to integrated departments, Hertin and Berkhout (2001) see a risk in that the environmental commitment and competence can be lowered and that the new order depends on the initial position of environmental issues on the agenda. Lafferty (2002) and Jordan (2002) argue that new institutions may not be necessary, but strong leadership from the core of government is. According to Lafferty, it is a necessary institutional factor to have an overarching authority structure, such as a chief executive, a planning agency, a body within the domain of the legislature or a last-resort judicial organ.

Keeping existing organizations intact, there is relatively strong support for various *accountability* mechanisms and the formal assignment of new *responsibilities or mandates* to promote EPI. Although the use of legal control and sanctions is still in its infancy, a responsibility incorporating the environment can be communicated in other ways (for example the Swedish sector responsibility initiative described in Chapter 1). The idea of this approach is to make sector departments internalize the policymaking principle of EPI. In this way they are forced to consider the sector's environmental impact and to develop their own environmental capacities in a decentralized and bottom–up fashion. Establishing formal accountability requires setting up an internal sector environmental monitoring and reporting mechanism (OECD, 2001a). In the longer term, changes of accountability and formal responsibilities may lead to an evolution of policy tradition and administrative culture, as new professional roles and tasks are developed. In the EU this approach has been implemented by instructing the sector council formations to develop their own strategies with targets and timetables under the Cardiff Process (see Chapter 1). Another accountability mechanism is to place a responsibility on an external organization to monitor and evaluate EPI progress.

Overcoming institutional fragmentation can also be facilitated by increasing *coordination and communication*, without necessarily changing organizational mandates or hierarchical relationships. Exploring how existing administrative coordination practices can be used for EPI has been a strong trend in recent EPI literature (see Meijers, 2004; EEA, 2005b; Schout et al, forthcoming). Interministerial committees and task forces can be formed, environmental correspondents can be sent to sector departments, and a central unit responsible for overview can be installed (OECD, 1996 and 2001b; Hertin and Berkhout, 2001). At the level of bureaucrats, networking schemes can be introduced, as well as regular circulation of staff between sector departments (OECD, 1996). In the UK there are two EPI initiatives of this kind: the Green Ministers Network and the Cabinet Committee on the Environment. However, success has been limited due to poor implementation, for example infrequent meetings (Jordan, 2002). Also, in the EU the success of these types of measures has been limited. Lebessis and Paterson (2000, in Hertin and Berkhout, 2001) have assessed measures such as environmental correspondents within each Directorate-General, incentives for staff rotation and the Green Star system to mark environmentally sensitive legislative proposals. They argue that:

> *Despite the clear need for the integration of policies and despite the increase in the use of tools designed to achieve it, there is evidence from the environment sector, for example, to suggest that there is frequently a tendency towards minimizing the possible influence of these tools. It appears that formal processes aimed at policy integration can become, in practice, little more than opportunities for representatives of different services to recite fixed positions.* (Lebessis and Paterson, 2000, in Hertin and Berkhout, 2001, p11)

Finally, *training and awareness* programmes in sector organizations are often part of proposed strategies for EPI. For example SEPA (2004) found that many smaller sector authorities do not have the expertise or skills to consider environmental objectives more systematically in their work.

Compared with procedural tools, organizational changes are likely to take a longer time to produce results. However, they may be more effective in causing lasting change than some 'add-on' procedural tools. Similar to procedural tools, however, changes of an organizational character also provide an infrastructure for EPI rather than necessarily solving the substantive issues and trade-offs related to it.

The normative approach

Compared with procedural and organizational tools, the normative approach directly addresses the substantive issues and trade-offs involved in EPI and thereby complements the construction of an EPI infrastructure. The underlying assumption here is that political will and priorities from the top down are the driving forces of sector policymaking. However, the success depends on the extent to which normative commitments and frameworks have high status and enjoy wide support, as opposed to being merely symbolic statements with limited critical engagement with the issues involved.

Most of the EPI literature emphasizes the need for *high-level political commitments* to make EPI a credible and active aspiration, as opposed to a principle on paper (see, for example, Lenschow, 2002e; Lafferty, 2002; OECD, 2002a). This commitment should also involve clear and strong leadership on EPI issues, in order to maintain political momentum. In addition to making EPI more prioritized further down in the governmental hierarchy, a clear commitment is also necessary in order to convey a democratic basis for EPI. However, in reality the links to concrete policy action and decision making in policy sectors are often unclear and the sincerity of EPI commitments can be questioned (Hertin and Berkhout,

2001). Politicians may be more interested in short-term 'electable' issues than developing long-term strategies for more diffuse, cross-cutting issues (Stead et al, 2004). Jordan (2002, p35) illustrates this problem in his analysis of EPI in the UK. According to Jordan, both 'hardware' (the organizations and procedures of governance) and 'software' (the knowledge needed to implement EPI) exist within UK government, which is known for 'having one of the strongest and most effective systems for coordinating departmental policies'. A main pillar of this system is the homogeneous and highly trained civil service. Still, EPI has not been successful due to the lack of political will, or the 'electricity' of the system. The lack of coordination and timing of EPI with political interests and attention cycles are identified as reasons for this.

A commonly identified means for ensuring effective translation of a commitment into concrete action is a formal overall *policy framework* for EPI, or sustainable development, in the government as a whole. Advising on general policy coherence, the OECD suggests that a comprehensive set of priorities should be established with a long-term view (OECD, 1996). Maintaining a strategic perspective is crucial, but also demanding in terms of information processing and potential political costs. In relation to EPI, a *national strategy for sustainable development* is commonly seen as such an overall framework (OECD, 2002a; Steurer and Martinuzzi, 2005). Based on an empirical examination, Lafferty and Meadowcroft (2000) found that with regards to horizontal EPI (HEPI), a national sustainable development strategy is extremely important, as it provides a platform for transcending difficult goal conflicts. However, formulating a policy framework for sustainable development also requires a clear definition of sustainable development, which can involve a long process in order to reach the 'common understanding' that the OECD (2002a) refers to. Lenschow (2002a) and Collier (1994) have both pointed towards the inherent ambiguities associated with sustainable development when drawing out more concrete policy priorities (see also Doern, 1993). Fundamental trade-offs involved in changing paths need to be addressed. Furthermore, it has to be realized that while overall benefits of EPI may be reaped at the national governmental level, there may be perceptions of 'net losses' at the sector levels in the short term. Fudge and Rowe's (2000) assessment of sustainability in Sweden suggests that these political challenges may be easier to overcome in societies characterized by homogeneity and a tradition of cooperation between the government and other actors.

While high-level political commitment can be thought of as 'pressure-from-above' for EPI, Lenschow (2002d, p243) argues that *societal backing* and public support is also necessary. She states that 'administrations rarely engage in path-breaking change unless they encounter pressure from the outside (crisis) or "below"' (ibid). Obviously, such a societal backing cannot be manufactured as part of an EPI strategy, but it still needs to be recognized as an important background condition.

As stated above, the normative approach towards EPI will be less effective if high-level commitments and policy frameworks are merely symbolic, since they will then fail to provide guidance and a source of legitimacy for EPI efforts in downstream policymaking. Another potential problem is if this top–down approach is not met with support, either from the administrative level or the general public.

The 'reframing' approach

The normative approach may thus lead to new environmental objectives and higher status of environmental concerns, but such top–down commitments may not 'stick' unless they enjoy wide political, administrative and public support. Another approach is to understand how more embedded and implicit sectoral ideas and discourses may lead to a revision of traditional sector objectives and reformulation of the sector policy rationale in the longer term. The question is if sector actors can 'reframe' their fundamental problem perceptions, causal narratives and overall policy goals into more environmental terms (Lenschow and Zito, 1998; Nilsson, 2005). For example, could transport sector policymakers replace the objective to provide a high, diverse and consistent supply of transport modes with one of facilitating mobility? In contrast with the normative approach, such a reframing process would involve more actors than top-level politicians and occur in a cumulative way, rather than simply issuing a new strategy. Revision of fundamental ideas could take place at both an individual level (attitudes, values, beliefs) and a contextual level (see Halpert, 1982, in Meijers, 2004). This approach is less common in the EPI literature, which has been relatively focused on developing practical tools for implementing EPI in the short and medium terms. Obviously, new ideas, discourses and frames cannot be forced on sector actors in the short term.

Nevertheless, Lenschow (2002c, p17) justifies the need to view this as an important analytical variable for EPI. Discussing the role of ideas, she argues that 'it is helpful to consider policy interests as embedded in a frame of reference, which prestructures the thinking within a policy sector'. Sustainable development is a new frame of reference and its acceptance depends on 'the relative persuasiveness of the causal story on the sectoral level'. Lenschow suggests that the general portrayal of sustainable development as a simple win–win scenario may make the idea seem less valid to policymakers. In a similar way, Nilsson (2005) has explored how different frames – with different understandings of EPI – have dominated over time in Swedish energy policy. On some occasions conceptual learning, or reframing, has occurred as opposed to mere technical learning. According to Collier (1994), the rare occasions of changing regulatory frameworks, for example liberalization of energy markets, offer excellent opportunities for creating structures and agendas for EPI.

Among the more concrete means suggested to make policy tradition and administrative culture more conducive to policy integration are research, training and socialization among decision makers. This 'intellectual strategy' proposed by Underdal (1980, p167) would lead to more comprehensive and holistic perspectives. Another reason for increased training is identified by OECD (2002b, p11) as the incorrect understanding (as a new name for environmental management) of the sustainable development concept among many civil servants. Larger awareness-raising campaigns and cultural shifts among the whole population have also been referred to as means for EPI (OECD, 2001a).

Thus this approach has been less explored in the EPI literature than the other ones, but it adds a longer-term perspective to the study of integration. Furthermore, it recognizes the importance of understanding how new ideas and commitments communicated from the top down are accommodated and internalized in the sectors (or not).

Conclusion: Towards a Better Understanding of EPI

Much like other broad concepts such as 'sustainability' and 'governance', there are difficulties involved in translating EPI – a value-laden, political principle with many connotations and interpretations – into a working concept for analytical study. The aim of this chapter has been to clarify some conceptual issues and provide an overview of the different perspectives on what EPI actually means and how it can be achieved. Despite many ambiguities, difficulties and contentions, EPI remains an important principle. Therefore, we need to know more about its practice and develop more effective measures to secure it.

Because the EPI challenge is multifaceted and operates on many different policy levels, a comprehensive approach is considered most useful. This means that the EPI principle has both a procedural and substantive dimension and should be studied in terms of process, output and impact. It should also recognize that the rationale for EPI can range from a purely organizational one to a strictly normative one. Most importantly, though, the review suggests that the tools-oriented EPI literature that predominates today needs to be complemented with a better understanding of the sector context within which EPI is supposed to occur. This involves understanding the existing sectoral ideas, institutions and policymaking processes and their evolution in the long term, rather than assuming that we can find discrete windows for integration where EPI tools can be applied with high receptiveness and effectiveness. Thus the research on generic EPI tools and the cross-country 'snapshots' of their use needs to be complemented with more research on mainstream sector policymaking, the target of integration efforts. Chapter 3 will present the approach taken in this book to address these needs.

With such a sector-context perspective on EPI, there are three particular gaps that need to be addressed. First, it is important to understand what kind of environmental problems should be addressed in sector EPI efforts, based on a comprehensive and systematic sector environmental assessment. Generally, little attention is paid to what the 'environmental' in EPI actually means and entails, especially if an add-on, procedural approach to EPI is taken. Which are the most important environmental impacts arising from sector activities? Are there internal environmental goal conflicts? A serious EPI effort clearly needs to be based on serious knowledge gathering. Despite the common recommendation to start EPI efforts with a sector environmental assessment, there appear to be few systematic methodologies for conducting them. A sector environmental assessment also brings attention to sector characteristics and problem characteristics, and their influence on achieving EPI. Determining the effectiveness of EPI tools for improving sector environmental impact with a causal analysis is likely to be difficult, but at least policy analyses can be contrasted with environmental impact trends.

Second, the weighting issue is central in EPI. However, rather than viewing it as an explicit judgement according to certain formal decision rules (such as 'principled priority' or the Pareto criterion) in a discrete decision-making situation, one could try to understand the sectors' embedded value systems behind implicit weighting. How have they incorporated environmental values and how have they evolved historically? Arguably, one gets a better idea of the 'big picture' in terms of EPI and what improvements have really been made in policy outputs by analysing sector policy change over a longer period of time and its relation to altered value systems, rather than by measuring the effect of new EPI tools in the short term. The understanding of such broader patterns is linked to the 'reframing' approach to EPI, which was found to have been only modestly explored. How do different policy frames held by key stakeholders understand the EPI challenge and how to tackle it? And how are these informed? At the most fundamental level, EPI is about reconceiving the key policy objectives of sectors in a way that makes the environment an intrinsic rationale of sector policy. There is a need for both methodologies and empirical case studies for understanding these processes.

Finally, rather than being primarily occupied with designated EPI tools that may be of limited relevance, we need to learn more about how the mainstream institutional framework, such as organizational and procedural arrangements for policymaking, are conducive or obstructive to EPI. An institutional analysis of sectoral 'mainstream' practices could address (if not provide a definite answer to) *how* environmental concerns move through the process and ultimately translate into different sector policy decisions. In addition to formal decision-making processes it also appears important to consider informal routines and activities of policy networks. Furthermore, if the study has prescriptive ambitions, it is important to pay close attention to the institutional dimension since it is difficult to make policy recommendations in relation to factors such as policy frames, actor configurations and power relations. Institutions can, at least partly, be consciously shaped and modified to accommodate new concerns, contexts and priorities.

References

Bina, O. (forthcoming) 'Strategic environmental assessment', in A. Jordan and A. Lenschow (eds) *Innovation in Environmental Policy? Integrating Environment for Sustainability*, Edward Elgar, Cheltenham, UK

Bomberg, E. (2004) 'Adapting form to function? From economic to sustainable development governance in the European Union', in W. Lafferty (ed) *Governance for Sustainable Development*, Edward Elgar, Cheltenham, UK

Collier, U. (1994) *Energy and Environment in the European Union*, Avebury, Aldershot, UK

Doern, G. B. (1993) 'From sectoral to macro green governance: The Canadian Department of the Environment as an aspiring central agency', *Governance*, vol 6, no 2, pp172–193

EC (2002) *Communication from the Commission on Impact Assessment*, COM(2002) 0276, European Commission, Brussels

Eggenberger, M. and Partidario, M. (2000) 'Development of a framework to assist the integration of environmental, social and economic issues in spatial planning', *Impact Assessment and Project Appraisal*, vol 18, no 3, pp201–207

EEA (2005a) *Environmental Policy Integration in Europe: State of Play and an Evaluation Framework*, European Environment Agency, Copenhagen

EEA (2005b) *Environmental Policy Integration in Europe: Administrative Culture and Practices*, European Environment Agency, Copenhagen

European Eco Forum (2003) *Environmental Policy Integration: Theory and Practice in the UNECE Region*, EEB, Brussels

Fudge, C. and Rowe, J. (2000) *Implementing Sustainable Futures in Sweden*, Byggforskningsrådet, Stockholm

Hertin, J. and Berkhout, F. (2001) 'Ecological modernisation and EU environmental policy integration', SPRU Electronic Working Paper Series, paper no 72, University of Sussex, Brighton, UK

Hertin, J. and Berkhout, F. (2003) 'Analysing institutional strategies for environmental policy integration: The case of EU enterprise policy', *Journal of Environmental Policy and Planning*, vol 5, no 1, pp39–56

Hey, C. (2002) 'Why does environmental policy integration fail? The case of environmental taxation for heavy goods vehicles', in A. Lenschow (ed) *Environmental Policy Integration: Greening Sectoral Policies in Europe*, Earthscan, London

Jacob, K. and Volkery, A. (2004) 'Institutions and instruments for government self-regulation: Environmental policy integration in a cross-country perspective', *Journal of Comparative Policy Analysis*, vol 6, no 3, pp291–309

Jordan, A. (2002) 'Efficient hardware and light green software: Environmental policy integration in the UK', in A. Lenschow (ed) *Environmental Policy Integration: Greening Sectoral Policies in Europe*, Earthscan, London

Lafferty, W. (2002) 'Adapting government practice to the goals of sustainable development', *Improving Governance for Sustainable Development*, OECD Seminar 22–23 November 2001, OECD, Paris

Lafferty, W. and Hovden, E. (2003) 'Environmental policy integration: Towards an analytical framework', *Environmental Politics*, vol 12, no 3, pp1–22

Lafferty, W. and Meadowcroft, J. (2000) *Implementing Sustainable Development: Strategies and Initiatives in High Consumption Societies*, Oxford University Press, Oxford, UK

Lenschow, A. (2002a) *Environmental Policy Integration: Greening Sectoral Policies in Europe*, Earthscan, London

Lenschow, A. (2002b) 'New regulatory approaches in "greening" EU Policies', *European Law Journal*, vol 8, no 1, pp19–37

Lenschow, A. (2002c) 'Greening the European Union: An introduction', in A. Lenschow (ed) *Environmental Policy Integration: Greening Sectoral Policies in Europe*, Earthscan, London

Lenschow, A. (2002d) '"Greening" the European Union – are there lessons to be learned for international environmental policy?', *Global Environmental Change*, vol 12, pp241–245

Lenschow, A. (2002e) 'Conclusion: What are the bottlenecks and where are the opportunities for greening the EU?', in A. Lenschow (ed) *Environmental Policy Integration: Greening Sectoral Policies in Europe*, Earthscan, London

Lenschow, A. and Zito, A. (1998) 'Blurring or shifting of policy frames? Institutionalization of the economic–environmental policy linkage in the European Community', *Governance*, vol 11, no 4, pp415–442

Liberatore, A. (1997) 'The integration of sustainable development objectives into EU policy-making: Barriers and prospects', in S. Baker, M. Kousis, D. Richardson and S. Young (eds) *The Politics of Sustainable Development: Theory, Policy and Practice within the European Union*, Routledge, London

Lundqvist, L. J. (2004) *Sweden and Ecological Governance: Straddling the Fence*, Manchester University Press, Manchester, UK

Meijers, E. (2004) 'Policy integration: A literature review', in D. Stead, H. Geerlings and E. Meijers (eds) *Policy Integration in Practice: The Integration of Land Use Planning, Transport and Environmental Policy-Making in Denmark, England and Germany*, Delft University Press, Delft, The Netherlands

Müller, E. (2002) 'Environmental policy integration as a political principle: The German case and the implications of European policy', in A. Lenschow (ed) *Environmental Policy Integration: Greening sectoral policies in Europe*, Earthscan, London

Nilsson, M. (2005) 'Learning, frames and environmental policy integration: The case of energy policy in Sweden', *Environment and Planning C: Government and Policy*, vol 23, pp207–226

Nilsson, M. (2006) 'The role of assessments and institutions for policy learning: A study on Swedish climate and nuclear policy formation', *Policy Sciences*, vol 38, pp225–249

OECD (1996) *Building Policy Coherence: Tools and Tensions*, OECD, Paris

OECD (2001a) *Policies to Enhance Sustainable Development*, OECD, Paris

OECD (2001b), *Sustainable Development: Critical Issues*, OECD, Paris

OECD (2002a) *Improving Policy Coherence and Integration for Sustainable Development: A Checklist*, OECD, Paris

OECD (2002b) *Governance for Sustainable Development: Five OECD Case Studies*, OECD, Paris

Persson, Å. (2004) *Environmental Policy Integration – An Introduction*, SEI, Stockholm

Peters, B. G. (1998) 'Managing horizontal government: The politics of coordination', Research Paper No 21, Canadian Centre for Management Development, Ottawa

Russel, D. and Jordan, A. (2004) 'Gearing up governance for sustainable development: Patterns of environmental policy appraisal in central government', Paper for the Political Studies Association Annual Meeting, 6–8 April 2004, Lincoln, UK

Schout, A., Jordan, A. and Unfried, M. (forthcoming) 'Administrative and bureaucratic mechanisms: The case of the European Union', in A. Jordan and A. Lenschow (eds) *Innovation in Environmental Policy? Integrating Environment for Sustainability*, Edward Elgar, Cheltenham, UK

SEPA (1999) *Samordning och målkonflikter: Sektorsintegreringens möjligheter och problem [Coordination and Goal Conflicts]*, Naturvårdsverket, Stockholm

SEPA (2000) Miljömål *och sektorsansvar [Environmental objectives and sector responsibility]*, Naturvårdsverket, Stockholm

SEPA (2003) *Society, Systems and Environmental Objectives: A Discussion of Synergies*, Conflicts and Ecological Sustainability, Naturvårdsverket, Stockholm

SEPA (2004) *Myndigheternas miljöansvar: Vidareutveckling av det särskilda sektorsansvaret [Agency Environmental Responsibility: Further Development of the Special Sector Responsibility]*, Naturvårdsverket, Stockholm

Stead, D., Geerlings, H. and Meijers, E. (2004) 'Working practices', in D. Stead, H. Geerlings and E. Meijers (eds) *Policy Integration in Practice: The Integration of Land Use Planning, Transport and Environmental Policy-Making in Denmark, England and Germany*, DUP Science, Delft, The Netherlands

Steurer, R. (forthcoming) 'Sustainable development strategies as policy integration processes: Coping with integrative governance challenges in Europe', in A. Jordan and A. Lenschow (eds) *Innovation in Environmental Policy? Integrating Environment for Sustainability*, Edward Elgar, Cheltenham, UK

Steurer, R. and Martinuzzi, A (2005) 'Towards a new pattern of strategy formation in the public sector: First experiences with national strategies for sustainable development in Europe', *Environment and Planning C: Government and Policy*, vol 23, no3, pp455–472

UK Cabinet Office (2000) 'Wiring it up: Whitehall's management of cross-cutting policies and services', *Performance and Innovation Unit Report*, January, Cabinet Office, London

Underdal, A. (1980) 'Integrated marine policy: What? Why? How?', *Marine Policy*, vol 1980, July, pp159–169

Volkery, A., Hertin, J. and Jacob, K. (forthcoming) 'Impact assessment and EPI: Procedures, trends and lessons', in A. Jordan and A. Lenschow (eds) *Innovation in Environmental Policy? Integrating Environment for Sustainability*, Edward Elgar, Cheltenham, UK

Weale, A. (2005) 'Governance and sustainability', paper presented at Governance of Sustainability: An International Research Conference, 23–24 June, UEA, Norwich, UK

Weale, A. and Williams, A. (1993) 'Between economy and ecology? The single market and the integration of environmental policy', in D. Judge (ed) *A Green Dimension for the European Community: Political Issues and Processes*, Frank Cass, London

3

Theory and Methodology
for EPI Analysis

*Åsa Gerger Swartling, Måns Nilsson,
Rebecka Engström and Lovisa Hagberg*

Chapter 1 formulated two overarching questions for the present study: first, how can one meaningfully interpret the principle of EPI in the real world and what is its current status and development in policy sectors? Second, what main factors affect EPI performance and how should institutions and knowledge-for-policy systems be shaped to advance EPI? Chapter 2 identified three gaps in previous research on EPI: the need for systematic assessments of sector environmental impact, exploration of EPI as a reframing process, and empirical analysis of institutional frameworks that might be conducive or obstructive to the EPI principle. This chapter presents how this book addresses these gaps from a methodological point of view and the basic theoretical underpinnings and methodology applied to studying EPI. It includes on the one hand studies of policy framing, learning and institutions, and on the other assessment of sectoral environmental performance. These supplement each other, but we also use them to make contrasting comparisons. For ease of exposition we describe only very briefly the essence of the applied theories and methodology. For complementary descriptions see Nilsson and Persson (2003), Nilsson (2005) and Engström et al (2006).

Conceptual Building Blocks

EPI as policy learning

As Chapter 2 showed in some detail, although EPI has been broadly embraced as an important principle for policymaking, different scholars have sought different explanations of how it can be understood and how it can be realistically achieved in policy processes. In this study, along with some recent work, we have endorsed a policy-learning approach to analysing EPI (Lenschow, 1999; Jepson, 2003; Nilsson, 2005). The approach echoes the growing attention to the role of learning in policy research, which is considered an important factor for policy innovation and policy change. Policy learning implies a change in thought about policy, which subsequently contributes to a change in the policy process. Policy change, in turn, is widely considered to be a prerequisite for sustainable development (Norberg-Bohn, 1999; Pearson et al, 2004).

The concept of policy learning originally related to the concept of cognitive dissonance in social psychology (Festinger, 1957). This refers to a tension that arises when a person is faced with inconsistent facts. Strategies to deal with this include a) modification of opinion in view of new facts or b) non-modification leading to strategies of denial or avoidance. As a general psychological phenomenon, the learning concept is applicable to processes at basically any level of society, from the individual to the international (Easterby-Smith and Lyles, 2003). The precise questions will obviously vary. In relation to policy learning the key question is how governments react to situations of cognitive dissonance. For instance, how do governments respond to new knowledge about environmental concerns and unsustainable development?

Learning approaches are manifest in many different fields including sociology, economics and policy analysis (Dierkes et al, 2001). They also exist within planning theory (Friedmann, 1996; Friedmann, 1987; Forester, 1993) and organizational theory (Argyris and Schön, 1978). The latter strand introduced the concept of 'single-loop' learning, which refers to learning that changes strategies or assumptions underlying strategies while the prevalent framework of values and norms remains unchanged. Single-loop learning is thus mainly preoccupied with effectiveness in achieving existing goals while maintaining organizational performance within the range specified by existing values and norms. This is contrasted with 'double-loop' learning, which results in a fundamental change in the values and norms of the applied theory and its strategies and assumptions at the individual or organizational level. In a similar vein to single-loop learning, 'technical' (Fiorino, 2001; Glasbergen, 1996) or 'instrumental' learning (May, 1992) are concerned with adjusting or modifying policy instruments in the context of stable policy objectives and strategies. Such learning occurs as part of a normally functioning policy system and does not entail fundamental reconsideration of the existing objectives. In contrast, 'conceptual learning', which is more rare, is, along with double-loop learning, concerned with changes in basic beliefs and paradigms. It entails redefining policy objectives and adjusting problem definitions and strategies to tackle policy problems. Studies have also brought attention to the role and nature of 'social learning' in policy and management for sustainability, which is conceptually close to double-loop learning and involves close interactions as well as a qualified dialogue among actors involved (Webler et al, 1995; Parson and Clark, 1995; Siebenhüner, 2004; Pahl-Wostl et al, 2003).

In his seminal work on social policy in Sweden and Britain, Heclo (1974) challenged the then widespread view that changes in public policy were largely a product of the clash between interests. Instead, he argued that gains and utilization of knowledge offered a better approach to explaining and comprehending how policy is formed. Today, however, most scholars would not view policy learning in opposition to interest politics. Instead, policy learning appears to be linked to, and can even be the result of, a process of strategic interaction of groups and organizations of elites lobbying for specific policy changes. Through his 'advocacy coalition framework', Sabatier (1988) explains how this may occur. Members of an advocacy coalition share a set of ideological and ontological beliefs about policy issues that shape policy positions, instrumental decisions and information sources selected to support specific policy positions. A number of factors can influence the coalition and its ability to achieve policy change.[1]

Notwithstanding the variance in the way learning is conceptualized and manifested in the policy arena, overall these learning approaches share the notion that learning is possible through adaptation to changing conditions and acquisition of new knowledge based on experience throughout the policy process.[2] While acknowledging the multitude of definitions and interpretations of EPI, this study has adopted policy learning as the key concept to understanding policy integration and policy change, defining EPI in terms of 'a policy-learning process in which perspectives evolve and reframe sectoral objectives, strategies and decision-making processes towards sustainable development' (Nilsson, 2005). EPI is therefore conceptually close to 'double-loop' and 'conceptual' learning but is concerned with sustainability aspects rather than learning more generally (Table 3.1).

Table 3.1 *Types of learning and EPI*

	Technical learning	**Conceptual learning**	
			EPI
Learning about	Instrument viability and effectiveness	Problem definitions, goals and strategies	Problem definitions, goals and strategies
Frames	Frames stable	Frames evolving	Frames evolving towards sustainability
Indicated in policy	Policy revisions of instruments or levels of instruments	Policy revisions of new problems and goals	Policy revisions of sustainability problems and goals
Indicated in argument	Accounts and citing of evaluations and experiences	New problem, goal and systems descriptions	Sustainability-led problem, goal and systems descriptions

Source: adapted from Nilsson (2005)

At the core of a learning approach is 'the notion that policy-makers and other actors can adjust to changing circumstances and to knowledge gained through experiences' (Fiorino, 2001, p330). A learning approach to EPI thus directs our attention to how knowledge on environmental issues is used in policymaking and to the institutionalization of knowledge over time. It has been shown that environmental concern goes through an 'issue-attention cycle', with rising and falling public attention (Downs, 1972). However, even though attention to environmental issues often decreases after a while, institutions, programmes and policies might have been introduced during the period of intensive attention, and these continue to work even after interest diminishes (ibid). Downs suggested that the general concern for issues that have gone through the attention cycle is almost always higher than those that are still in the pre-discovery stage, which means there is a certain amount of institutionalization of knowledge in the system.

EPI in the policy network

Who learns? In the literature, policy integration is sometimes portrayed as an administrative process occurring in hierarchical and goal-oriented governmental structures where governments establish political preferences and control outcomes. From such a perspective, policy integration arguably depends on 'rationalistic' criteria such as ensuring comprehensiveness in inputs, aggregation in preferences, and consistency between implementation and intentions (Underdal, 1980). However, such perspectives have been criticized for ignoring situations of multiple actors with distributed powers and competing interests (Hogwood and Gunn, 1984). In contrast to these hierarchical perspectives of policy formation, our approach to policy learning is based on the premise, captured in the policy-network theory, that policy is formed in networking processes with multiple actors both public and private (including government organizations, interest groups, political parties and scientists) that have different ideas and interests (de Bruijn and ten Heuvelhof, 2000; Peters, 1998). Policy networks can be understood as informal structures for communication and interaction between actors. Extending beyond formally institutionalized decision makers in government, they include many interconnected actors, each of which draws on particular resources to influence the way public and private policies are shaped and implemented. In adopting a network perspective on policymaking, as opposed to the more hierarchist model of (top–down) policymaking, we consider ideas, interests and knowledge occuring broadly in society, considering actors and actor coalitions from interest groups, political parties, businesses and academia as well as the governmental system.

It has been argued that where policies do not change much it is because policy communities are able to control decision making. In his examination of nuclear policies in France and the US, Baumgartner (1989) shows how the arrangement between the state and the interest group has a decisive influence on the political process. Changing the configuration of participants in a process is an effective way of changing the balance of power on an issue. Whether such a strategy will be successful depends on the legislative arrangements, the networks involved and their roles. However, the complex structures of contemporary policy networks makes the analysis challenging. Not only are there multiple actors and multi-level political systems involved in the shaping of policy, but the actors in the political arena influence, and are influenced by, each other as well as by external political events and processes. Ideas and interests are moving targets, shaped and formed by interactions and processes within the network. Capturing this dynamic is at the core of a policy-learning study. Below we describe how EPI as policy learning is identified through analyses of policy framing in the policy formation processes at the national level over time.

Learning as changes in policy framing over time

What is learned? In our view, EPI is about adding environmental values and perspectives to the ways in which a sector considers policy issues, as policymakers and actors adjust their combinations of means and objectives with respect to experience and new knowledge. This implies that learning has not only occurred as an ad hoc application of environmental policy instruments due to demands which have their origins outside the sector, but can be portrayed, as explained above, as a special case of conceptual learning towards sustainable development. Such learning is not easily captured empirically, as evidenced by the rather meagre empirical literature since the concept emerged in the 1970s. In this study, it is suggested that it can be observed by combining observations of changes in the actors' problem understandings, rationalities and argumentations with studies of more direct objectives and strategies (May, 1992; Hall, 1993; Sabatier, 1988). However, to avoid a fragmented and disjoined analysis, it is useful to introduce a unifying concept. The evolution of *policy framing*, a concept that has been deployed in a few EPI studies before (see Chapter 2), is one way of creating a coherent perspective.

Policy frames have been defined as 'ways of selecting, organizing, interpreting and making sense of a complex reality to provide guideposts for knowing, analysing, persuading and acting' (Rein and Schön, 1993).[3] Frames can be regarded as consisting of beliefs, values and perspectives that help actors to derive coherence and meaning out of thinking and action in policy matters. Policy frames contain both objectives, causal assumptions about problems, and prescriptions about suitable responses. They remain fixed in the context of technical learning, or single-loop learning, but evolve under conceptual learning through, for example, frame reflection, alignment and bridging, and reframing (Benford and Snow, 2000; Schön and Rein, 1994). Frames are usually tacit. This implies that, in order to become aware of different frames that may underlie policy controversy, they must be constructed both from the texts and speeches in the debate preceding decisions and from the actual decisions, regulations and routines set up to sustain a policy. This is difficult: a particular action may also be coherent with different frames, or conversely, a frame may result in different courses of action.

Frames thus evolve as a result of conceptual learning. However, it should be noted that policy frames and reframing are not only influenced by policy-learning processes but also factors such as politics and strategic behaviour. Actors can, to some extent, construct new frames to advance their positions, for instance through reframing an environmental protection issue into a technology innovation one. This was quite common under the ecological-modernization strategies pursued in the 1990s (Hajer, 1995). In this way, competing frames can start to speak to each other, which in turn can induce learning across frames, allowing better coordination and treatment of trade-offs between, for instance, market reform and environmental interests. Such processes are identified in Chapter 5.

Finally, in any policy sector, there are many parallel and sometimes overlapping policy processes. Some of these occur over a long time period, whereas others are short and intensive. Although it would appear logical to study framing in the early processes of policy formation, and at the level of arguments and discourses, it has been argued that the framing of problems should also be studied at later stages of the policy process (Mazey and Richardson, 1997).[4] If not in the actual implementation of policy, frames should at least be sought at those stages where policy is specified in terms of choice of policy instruments and overall strategy. In empirical terms, this requires studying not only arguments and rhetoric but also policy output.

Contextual and institutional factors influencing EPI

Drawing on the previous discussions on how EPI can be identified and traced, we now move on to consider what causes or hinders EPI. Chapter 2 concluded that EPI studies need to pay particular attention to the sectoral context, including institutional arrangements. What conditions may affect EPI? The policy-learning literature points, in particular, towards new societal problems, domestic politics, international policy streams, disasters or economic crises as stimuli for learning and frame adjustment (Fiorino, 2001; Weale et al, 2000; Sabatier, 1988). Rose (1991) refers to the learning process as one in which policymakers draw lessons in response to dissatisfaction with the implications of, for example, globalization trends or budgetary problems. This state of dissatisfaction stimulates the search for new solutions in the policy system.

Most scholars agree that the potential for learning and knowledge use in policymaking is partly a function of the issue characteristics, including aspects such as knowledge level, goal conflicts and synergies. It has been argued that the impact of scientific knowledge is likely to be strong when there is 'definite' or at least consensual knowledge on the problem; a feasible 'cure' can be presented; effects are close in time and space; the problem is affecting the 'social centre' of society; the problem is developing rapidly and surprisingly; the effects are experienced by or at least visible to the public; the political conflict is low; and there is an institutionalized setting and iterative decision-making (Underdal, 1989). The level of political conflict can be particularly salient, as it also influences the policymakers' incentive to reduce or expand the issue to the general political arena, which in turn determines the extent to which it becomes a political, rather than a technical issue (Baumgartner, 1989). The implications of this politicization are not clear. It highlights *political will* as another important factor for EPI, and it is commonly argued that if there is no real commitment to EPI among decision makers, it is likely to remain a decorative flourish in policy documents that has little or no bearing on policy outcomes (Hertin and Berkhout, 2003). However, political will is a Janus-faced phenomenon, as much a driver as a result of policy learning and EPI.

Throughout the empirical analysis, we endeavour to keep track of these larger contextual aspects. Also, our analytical approach is to investigate how *institutions* of policymaking affect the possibility of achieving EPI in the sectors. Institutionalism has become an influential approach to analysing political life and policy decisions generally. It emphasizes the role that formal and informal rules have in shaping and constraining, if not determining, policy outcomes (Scharpf, 1989). As opposed to older traditions of institutional analysis, new institutionalism implies a shift from formal structures of governance to informal behavioural-oriented ones; today, institutions are commonly considered to include *rules* that allocate authorities, *norms* that govern the preferences of various actors and *practices* that govern behaviour (Weale et al, 2000). Peters (2000) identifies three key features as the defining elements of institutions. First, they are formal or informal structures in society (including a law, an agency, a network of organizations or a shared set of values). Second, there is some stability over time, although some might be more stable than others. Third, they affect individual behaviour, and can therefore be seen as constraints.

The institutional framework shapes the flow of ideas, the construction of interests, the nature of power relations, and the form of interaction between (competing) actors and interests. That lack of EPI can be seen as an institutional problem has been confirmed through numerous previous studies (Lenschow, 2002; Connelly and Smith, 1999; Jordan, 2002; Nilsson and Persson, 2003). At the policy level, important institutions include distribution of responsibility and resources and how the policy process is organized, as well as policymaking procedures, such as who has access to decision making, in what manner different actors are included, what procedures for knowledge are used and by what criteria decisions are taken. In particular, coordination systems, planning and monitoring systems, assessment procedures, and consultation procedures have been pointed out as important factors in achieving EPI (Jordan and Schout, 2006). Hence institutions could be viewed as an intermediate variable that enables, or prevents, the integration of environmental concerns in sector policies. Ultimately, however, we deploy an evolutionary perspective on institutions. In this lies the stickiness of institutions, along with the fact that they change over relatively short time periods. Recent administrative reforms have partly been influenced by external international political doctrines and events such as crises or initiatives, but also by internal historical-institutional factors, culture, tradition and style (Weale et al, 2000). This perspective is a middle ground between the optimistic view of the possibility for full comprehensive reform and a more negative view that institutional reform is impossible. By examining the institutional conditions that seem to favour EPI, we will also seek to develop recommendations for how institutional arrangements can be made to better support EPI.

Empirical Approach and Methods

The concepts presented above are analysed and compared across the case studies as well as over time. Learning studies generally require long time frames, at least over a decade (Sabatier, 1993). Since the period of study is too long to allow for a continuous in-depth analysis of all the ongoing policy processes within the two case study sectors, we have chosen to focus on a few selected 'policy rounds' (Teisman, 1992). Policy rounds are identifiable processes in which a certain set of issues is treated over a longer or shorter period of time. In theory, they comprise the typical procedures characteristic of a policy process: problem identification, identification of possible solutions, policy decision and implementation. A round may involve what would typically be considered a number of processes that converge around the same theme, for example appointment of committees, negotiations between political parties and preparations in ministries. In Chapters 5 and 6, the policy rounds studied converge around major policy decisions from the late 1980s, mid 1990s and early 2000s and typically last around two to five years. This generates an empirical account of how the studied sectors organize themselves and how their institutional procedures have evolved as new policies have been prepared and policy frames have changed. To systematize the institutional data collection for Chapter 6, we use a set of common questions.

Numerous primary sources of documentary data are used, including government bills, committee reports, government communications, protocols from the parliament, reports produced by sector authorities, and policy documents, articles and debates by various interest groups. These documents are also used in order to trace the institutional context in which relevant processes have occurred. Policy output in the form of decisions and regulations are used to identify substantial learning, for example in terms of the adoption of new policy instruments. Secondary sources are used in order to trace the history of policy processes and to get deeper knowledge of the sectors.[5]

Interviews are a crucial part of the methodology. They were used extensively in order to learn more about policy processes in the respective sectors, especially the decision procedures and the use of knowledge. Interviews have appeared to be particularly useful with respect to understanding individual perspectives, experiences, strategic behaviour and intentions among stakeholders that are not possible to obtain through documents. A number of pilot interviews were carried out in 2003 in order to get a general overview of the state of EPI in each sector, before the analytical framework was developed. Interviewees include policymakers and other actors engaged in the national policy arena, including members of parliament, officials from relevant ministries, governmental agencies, and representatives of the business sector, the scientific community and NGOs. Interviewees were selected on the basis of the written material and through the 'snowball method', whereby interviewees were asked to identify other persons whom it would be valuable to talk to. A total number of approximately 60 interviews were conducted between 2003 and 2006, all of which were fully transcribed and analysed.

Sector environmental analysis

Apart from the analysis of EPI in policy processes, the study has developed and applied a methodology for studying the environmental outcomes in the agricultural and energy sectors. It would be extremely difficult, if not impossible, to causally relate this sector environmental impact to the learning and EPI processes that take place at the national policy level. This has not been a purpose of the methodology. Instead, the aim is to get a better grip on and understand more about the 'ultimate raison d'être' of EPI, in other words the environmental outcomes in the two sectors, and to evaluate which of these are the most serious for each sector and how impacts can be reduced.

The sector environmental analysis employs a combination of methods for environmental systems analysis including the tools 'environmentally extended input–output analysis' (IOA) and 'life-cycle assessment' (LCA). These different tools are useful for answering different questions; combined they provide a more comprehensive analysis. The tools share a systems view of the object of study, meaning that upstream and downstream effects are taken into account. Using the systems view leads to a somewhat different definition of the sectors compared to the one used in other parts of this book. The starting point is the sectoral production in one year, and the impacts arising from this production. Using the agriculture sector as an example, the analysis covers both impacts from the agriculture itself, including impacts from producing all inputs needed on the farms, such as machinery, pesticides and fertilizers, as well as impacts from processing, transport and consumption of all products from agriculture. This definition might at first seem unnecessarily wide. However, to strive for effective resource use with low levels of pollution, it is important to include all upstream and downstream effects. Otherwise the problems might be moved up- or downstream in the production chain, rather than being truly reduced. For example, if Swedish farmers fed their livestock with imported manufactured feed, rather than on crops from domestic farms, environmental impacts would appear to be decreasing in a study limited to Swedish farms. However, in reality they would only be shifted to other parts of the world. With our methodology, such effects are covered.

The method applied for the sector analysis can be seen as a hybrid IOA-LCA study (Suh and Huppes, 2005). To find out the magnitude of each problem in the sectors we use environmentally extended IOA as a base. However, there are environmental effects from the sectors not captured in the IOA, hence the analysis is complemented with data from other sources, such as LCAs of products from the sectors. The national environmental quality objectives are used as a checklist to capture relevant impacts. The methods of IOA and LCA are briefly described below. For a more comprehensive discussion of the methodology the reader is referred to Engström and Wadeskog (2006).

IOA is a well-established analytical tool within economics and systems of national accounts using a nation or a region as the object of the study (Miller and Blair, 1985; UN, 1999). The input–output matrices describe trade between producers and users. IOA can be applied to include environmental impacts either by adding emission coefficients to the monetary IOA or by replacing the monetary input–output matrices with matrices based on physical flows. The former is the type most often used and discussed (Lave et al, 1995; Joshi, 2000). The environmentally extended IOA focuses on the environmental pressures the production causes in terms of energy use, emissions to air and water, and use of chemicals. This is done using environmental data allocated by industry and therefore linked directly to the production levels of the different industries. The analysis provides information on resource use and emissions per monetary unit for each sub-sector, for example carbon dioxide (CO_2) per US\$. Environmental data are collected in a satellite accounting system linked directly to the national accounts. A framework for such environmental accounts has been developed by the UN (1993 and 2003) and such systems are established in many countries. The method we use for the sector analysis will therefore also be applicable outside Sweden.

LCA is a tool to assess the environmental impacts and resources used throughout a product's life from raw material acquisition through production use to disposal. The term 'product' can include service systems as well as product systems. The assessment is standardized in the ISO 14040 series (ISO, 1997, 1998 and 1999), and a guide to the standards has been developed (Guinée et al, 2002). Apart from an inventory, LCA includes methodologies for classifying into impact categories, for example global warming or human toxicity. After that, an optional step is to employ methods of normalization, grouping and weighting in order to evaluate which are the most important environmental impacts. Several methodologies exist, and initiatives have been taken to develop best practice (Udo de Haes et al, 2002).

In order to evaluate which problems are the most urgent for each sector, in the first step we aggregate data from the IOA-LCA method in impact categories. The following categories are applied in this context: non-renewable resources, global warming, human toxicity, freshwater toxicity, marine water toxicity, terrestrial toxicity, eutrophication, acidification and photochemical oxidation. The baseline characterization developed at Leiden University is used (Guinée et al, 2002) as included in the SimaPro 5.0 (Goedkoop and Oele, 2001), with the exception of non-renewable resources, where we use a thermodynamic approach (Finnveden and Östlund, 1997).

In the next step, the impact categories are weighted using three different methods: Ecotax 2002 (Finnveden et al, 2006), Ecoindicator 99 (Goedkoop and Spriensma, 2000) and EPS 2000 (Steen, 1999), the latter two as implemented in the SimaPro software (Goedkoop and Oele, 2001). The identification of the most important problems becomes more robust than if using only one of these methods. Ecotax employs weighting factors derived from Swedish ecotaxes as a measure of how Swedish society values different impacts. Ecoindicator applies damage models that link the data on emissions and resource use to three different damage categories – human health, ecosystem quality and resources – which are weighted using a panel approach. EPS evaluates impacts on the environment using willingness-to-pay measures. Some impacts, such as biodiversity, could not be quantified in such a way that they could be incorporated into the characterization and weighting methods; therefore we also compare the impact from each sector with Sweden's total impact in order to evaluate the relative contribution to the problem.

Framing analysis

Having gone through the sector environmental analysis in Chapter 4, the book proceeds to perform a policy-framing analysis to detect patterns of conceptual learning and EPI over time in Chapter 5. Through a systematic tracing of agendas, argumentation and policy evolution we identify and investigate important framing and reframing patterns in the energy and agricultural policy sectors. The frames are identifiable on the basis of a comprehensive documentary study using qualitative and quantitative content analysis techniques, complemented with and cross-checked against an overall analysis of transcripts from the interviews with stakeholders engaged in the Swedish national policy domain.

Content analysis can be broadly defined as 'any technique for making inferences by objectively and systematically specified characteristics of messages' (Holsti, 1969, p14). It has also been more precisely defined as the systematic, replicable technique for compressing many words of text into fewer content categories based on explicit rules of coding (Krippendorf, 1980; Berelson, 1952; Weber, 1990). There are, in general, several reasons to consider applying content analysis in textual data collection and management. For example, Stemler (2001) contends that content analysis enables the researcher to go through large volumes of data with relative ease in a systematic fashion. Referring to Weber (1990) he further argues that it can be a helpful tool for allowing us to discover and describe the focus of individual, group, institutional or social attention.

There are several advantages of adopting a content-analytical approach for textual management and analysis of our two case studies. Content analysis is appropriate since we are interested in identifying critical patterns and trends over a (longer) time period in large amounts of text. It may therefore help us identify critical policy rounds over a longer time period before we enter into a more exploratory analysis stage where idea analysis becomes a more appropriate tool. It enables us to go through the large volume of documents relating to energy and agricultural policy issues over a 15–20 year period in a systematic fashion. It also provides an empirical basis for monitoring shifts in the contemporary debate on energy and agricultural policies in Sweden. It enables us to compare data over the selected period to determine how policy changes have occurred and manifested themselves in, for example, government documents. Content analysis is advantageously combined with a more interpretive and exploratory text-analytical method, and several authors advocate a multi-method approach to reflect the multidimensionality of public policy issues (McQuail, 1992; Majchrzak, 1984).

While the traditional content analysis is mainly about measuring and quantifying key data in texts (see, for example, Holsti, 1969 or Krippendorf, 1980), qualitative content analysis has been developed to become a useful tool for a rule-guided systematic approach to preserve some of the methodological strengths of quantitative content analysis and widen them to a concept of qualitative procedure (Mayring, 2001). Thus qualitative content analysis aims at combining the quantitative content analysis approach for systematic analysis of large amounts of textual material with a qualitative-oriented procedure of text interpretation in a manner that is suitable and applicable to our study. With an emphasis on qualitative rather than quantitative analysis, one may be a free interpreter of the material, with content analytical steps and rules only as orientation, while establishing a subjective relation to the material (Mayring, 2001).

Understanding policy framing requires a more interpretative approach than a pure content analysis can offer because languages as well as ideas are important in understanding frames. It is also important to recognize the active manipulation of frames by actors: although actors are partly constrained by frames, they can also use frames in a conscious manner. However, frames are still viewed as more fundamental than actors in explaining a policy process: framing is not just something that actors do, but also something that constitutes the world of actors, and thus their perceived policy options. This also suggests that actors, depending on the context, can move between frames and be quite flexible in their use of frames. The frames can be used strategically to favour certain interests and are thus heavily informed by politics. EPI as a learning process is then understood as the relationship between the different frames, allowing for adjustment through coordination and trade-off between the perceived policy options.

Two ways to carry out idea analysis can be distinguished (Bergström and Boréus, 2000). The first method is to identify ideal types (for example of frames) in order to analyse what kind of ideas or ideology is represented in a certain document. In our empirical study, the advantage of this would be a stringent analytical model according to which frames could be sorted. On the other hand, we would have to construct ideal types of frames for every case, which would take considerable time. Furthermore, ideal types would in themselves represent a kind of result and thus risk introducing a circular argument to the analysis (Bergström and Boréus, 2000, p170). Another method, where the ideal types are not known, is to identify important dimensions or themes on which texts are supposed to differ. Dimensions could be described as dichotomies (for example, the individualist–collectivist dimension in welfare politics). Whereas it is certainly possible to argue that some issues could be thought of in this way (for example preferences for market versus political regulation in the agriculture and energy sectors), others are not easily fitted into such a structure. Therefore a thematic approach (with themes grounded in the literature on frames and EPI) seems more appropriate. The themes were addressed through a joint set of questions, which should help answer both how the current situation within the sector is framed (challenges, opportunities, conflicts, etc) and ideas about in what direction it should be developed. The questions, used to analyse both documents and interview transcripts, are at a rather general level and are further specified in relation to each case (Table 3.2).

Table 3.2 *Framing questions for agriculture*

Theme 1 Agriculture sector characteristics

What main problems does agriculture face today?

What should be done about them?

What opportunities are there to transform agriculture?

What is agriculture's contribution to society? Should it change, and if so, how?

What is the role of the farmer? Should it change, and if so, how?

What measures are or should be used to evaluate success in the sector?

Which actors belong to the sector?

What is the role of the sector in relation to other sectors or policy areas?

Theme 2 The agricultural sector and ecological sustainability

Is there potential for the sector to become sustainable?

What are the main opportunities and obstacles for achieving sustainability within the sector?

What is it that should be sustainable?

At what scale should sustainability be evaluated?

Within what time frames should sustainability be evaluated?

What, if any, is the role of agriculture in moving towards a sustainable society?

Theme 3 Governing agriculture

How is sector governance organized, and how should it change?

In what proportions should the political sphere and the market be used in order to govern the sector?

What kinds of instruments are best used to govern the sector?

What should be the relation between expert and public input to policy?

Conclusions

This book's analysis of EPI in sectors combines several methodologies. At the most general level, we conceptualize EPI as a process of policy learning in which sectoral policy perspectives evolve and reframe objectives, strategies and decision-making processes towards sustainable development. To study the outcomes of the EPI process, we thus analyse the reframing of policy over time. A relatively long time period is needed to capture these changes: our range is from the late 1980s to the mid 2000s. This policy-framing and learning analysis, reported on in Chapter 5, addresses both the policy instruments and strategies in place and the underlying argumentation and reasoning. In each case study, we zoom in on three policy rounds distributed over time and identify patterns of policy framing and EPI. This focuses on signs and attributes of learning and frame adjustments manifested in relevant policy processes and outputs that appear to promote environmental considerations in sector policy. We also carry out a quantitative content analysis of major policy bills to trace the emergence of the different environmental issues in the policy bills. In Chapter 4 we also investigate the environmental impacts of the sectors to characterize their ultimate environmental performance and which issues need to be considered.

Due attention is paid to causal relationships that have influenced the observed learning and framing patterns, including both external and internal factors. It is recognized that the causal factors relating to EPI operate in complex, dynamic settings which are difficult to unpack fully. Our ambition is to analyse some major causal links, although we do not endeavour to clarify exactly how and in which hierarchical order all factors interact. In Chapter 5 we study the broader contextual factors that have influenced reframing and EPI, while in Chapter 6 we focus on how institutions have affected EPI. Ultimately, the study of institutional factors aims to identify what types of institutional arrangements appear to be particularly conducive to EPI. To understand what main factors affect EPI performance and how institutions may advance, we use qualitative content analysis of policy documents, as well as interviews with key informants in the sectors.

Notes

1   'Stable influences' are, for example, established policy parameters and the social, legal and resource dimensions of the society that persist over a period of several decades; 'dynamic influences' for example external changes or activities in global socio-economic conditions that can alter the composition and resources of various coalitions. Personnel changes at senior levels within government ministries can also affect the political resources of various coalitions and the decisions that are made at the collective policy choice and operational levels.

2   According to this notion, the policy learning approach is clearly different from the assumptions made by Rational Choice Theory (RCT), which departs from the perspective of the individual and his or her interests and rational choices, rather than from several individuals interacting together and from social situations. In comparison with RCT assumptions, learning is more stable over time, as seen in its reference to the way in which, as Heclo (1974) has noted, a relatively enduring alteration in policy results from policymakers' and participants' learning from their own, and others', experience with similar policies.

3   Conceived in the 1970s, policy frames was popularized in political science and political sociology in the 1990s (Goffman, 1974; Schön and Rein, 1994). They emerged in an ideational tradition

of political sciences that contains concepts such as paradigms (Hall, 1993), belief systems (Sabatier, 1988), values (Dunn, 1994) and discourses (Hajer, 1995; Dryzek, 1997). Although these concepts vary, they all embrace the notion that policy change requires an evolution of perceptions and interpretations of reality that give meaning to political preferences and arguments.

4   Mazey and Richardson (1997) hold that, whereas frames may be rhetorical and therefore of passing importance in policymaking as actors learn to espouse the language of the most dominant frames, it might be more difficult to overcome the difference between frames in practice, in other words in implementation. According to this argument, it is not necessarily when ideas are on the table that they will cause conflict, but when they are shrouded in the more down-to-earth concerns of how to make them real.

5   In general, there is more material available on energy than on agriculture in Sweden. For example, there is a myriad of energy policy documents ranging from government bills, government communications and committee reports relating to energy, nuclear and climate policies. All government publications and reports from 1994 are available for downloading on the government's website (www.regeringen.se). There are also a number of books available on Swedish energy policy. Whereas the institutional context of Swedish agriculture has received comparatively limited attention, there is a large literature on the EU Common Agricultural Policy and its history (Kay, 1998).

References

Argyris, C. and Schön, D. (1978) *Organizational Learning: A Theory of Action Perspective*, Addison-Wesley, Reading, MA, US

Baumgartner, F. (1989) 'Independent and politicized policy communities: Education and nuclear energy in France and in the United States', *Governance*, vol 2, pp42–66

Benford, R. and Snow, D. (2000) 'Framing processes and social movements: An overview and assessment', *Annual Review of Sociology*, vol 26, pp11–39

Berelson (1952) *Content Analysis in Communication Research*, Free Press, Glencoe, UK

Bergström, G. and Boréus, K. (2000) *Textens mening och makt. Metod i samhällsvetenskaplig textanalys [The Meaning and Power of the Text. Method in Social Scientific Text Analysis]*, Studentlitteratur, Lund, Sweden

Connelly, J. and Smith, G. (1999) *Politics and the Environment: From Theory to Practice*, Routledge, London

de Bruijn, H. and ten Heuvelhof, E. (2000) *Networks and Decision Making*, LEMMA, Utrecht, The Netherlands

Dierkes, M., Antal, A. B., Child, J. and Nonaka, I. (eds) (2001) *Handbook of Organizational Learning and Knowledge*, Oxford University Press, Oxford, UK

Downs, A. (1972) 'Up and down with ecology: The "issue-attention cycle"', *The Public Interest*, vol 28, pp38–50

Dryzek, J. S. (1997) *The Politics of the Earth*, Oxford University Press, Oxford, UK

Dunn, W. N. (1994) *Public Policy Analysis – An Introduction*, Prentice-Hall, Englewood Cliffs, NJ, US

Easterby-Smith, M. and Lyles, M. A. (2003) *The Blackwell Handbook of Organizational Learning and Knowledge Management*, Blackwell, Malden, UK

Engström, R. and Wadeskog, A. (2006) 'Environmental impact from a sector: Production and consumption of energy carriers in Sweden', submitted manuscript

Engström, R., Wadeskog, A., and Finnveden, G. (2006) 'Environmental assessment of Swedish agriculture', *Ecological Economics,* available online

Festinger, L. (1957) *A Theory of Cognitive Dissonance*, Stanford University Press, Stanford, US

Finnveden, G. and Östlund, P. (1997) 'Exergies of natural resources in life-cycle assessment and other applications', *Energy*, vol 22, pp923–931

Finnveden, G., Eldh, P. and Johansson, J. (2006) 'Weighting in LCA based on ecotaxes. Development of a mid point method and experiences from case studies', *International Journal of LCA*, no 11, pp81–88

Fiorino, D. (2001) 'Environmental policy as learning: A new view of an old landscape', *Public Administration Review*, vol 61, pp322–334

Forester, J. (1993) 'Learning from practice stories: The priority of practical judgment', in F. Fischer and J. Forester (eds) *The Argumentative Turn in Policy Analysis and Planning*, Duke University Press, London

Friedmann, J. (1996) 'Two centuries of planning theory: An overview', in S. J. Mandelbaum, L. Mazza and R. W. Burchell (eds) *Explorations in Planning Theory*, Centre for Urban Policy Research, New Brunswick, Canada

Friedmann, J. (1987) *Planning in the Public Domain: From Knowledge to Action*, Princeton University Press, Princeton, US

Glasbergen, P. (1996) 'Learning to manage the environment', in W. Lafferty and J. Meadowcroft (eds) *Democracy and the Environment: Problems and Prospects*, Edward Elgar, Cheltenham, UK

Goedkoop, M. and Oele, M. (2001) *User Manual. Introduction into LCA Methodology and Practice with SimaPro 5*, Pré Consultants, Amersfoort, The Netherlands

Goedkoop, M. and Spriensma, R. (2000) *The Eco-Indicator 99: A Damage Oriented Method for Life Cycle Impact Assessment*, Pré Consultants, Amersfoort, The Netherlands

Goffman, E. (1974) *Frame Analysis: An Essay on the Organization of Experience*, Harvard University Press, Cambridge, US

Guinée, J. B., Gorrée, M., Heijungs, R., Huppes, G., Klejn, R., Koning, A., van de Oers, L., Wegener Sleeswijk, A., Suh, S., Udo de Haes, H. A., Bruijn, H., van Duin, R. and Huijbregts, M. A. J. (2002) *Life Cycle Assessment, An Operational Guide to the ISO Standards*, Kluwer Academic Publishers, Dordrecht, The Netherlands

Hajer, M. (1995) *The Politics of Environmental Discourse: Ecological Modernization and the Policy Process*, Oxford University Press, Oxford, UK

Hall, P. (1993) 'Policy paradigms, social learning, and the state: The case of economic policymaking in Britain', *Comparative Politics*, vol 25, pp275–296

Heclo, H. (1974) *Modern Social Politics in Britain and Sweden: From Relief to Income Maintenance*, Yale University Press, New Haven, US

Hertin, J. and Berkhout, F. (2003) 'Analysing institutional strategies for environmental policy integration: The case of EU enterprise policy', *Journal of Environmental Policy & Planning*, vol 5, pp39–56

Hogwood, B. and Gunn, L. (1984) *Policy Analysis for the Real World*, Oxford University Press, Oxford, UK

Holsti, O. R. (1969) *Content Analysis for Social Sciences and Humanities*, Addison-Wesley, Reading, MA, US

ISO (1997) *Environmental Management – Life Cycle Assessment – Principles and Framework*, International Standards Organisation, Geneva, Switzerland

ISO (1998) *Environmental Management – Life Cycle Assessment – Goal and Scope Definition and Inventory Analysis*, International Standards Organisation, Geneva, Switzerland

ISO (1999) *Environmental Management – Life Cycle Assessment – Life Cycle Impact Assessment*, International Standards Organisation, Geneva, Switzerland

Jepson, E. J. (2003) 'The conceptual integration of planning and sustainability: An investigation of planners in the United States', *Environment and Planning C: Government and Policy*, vol 21, pp389–410

Jordan, A. (2002) *Environmental Policy in the European Union: Actors, Institutions and Processes*, Earthscan, London

Jordan, A. and Schout, A. (2006) *The Coordination of the European Union*, Oxford University Press, Oxford, UK

Joshi, S. (2000) 'Product environmental life-cycle assessment using input-output techniques', *Journal of Industrial Ecology*, no 3, pp95–120

Kay, A. (1998) *The Reform of the Common Agricultural Policy: The Case of the MacSharry Reforms*, CABI Publishing, Wallingford, UK

Krippendorf (1980) *Content Analysis*, Sage, London

Lave, L. B., Cobas-Flores, E., Hendrickson, C. T. and McMichael, F. C. (1995) 'Using input–output analysis to estimate economy-wide discharges', *Environmental Science and Technology*, vol 29, pp420A–426A

Lenschow, A. (1999) 'The greening of the EU: The Common Agricultural Policy and the Structural Funds', *Environment and Planning C: Government and Policy*, vol 17, pp91–108

Lenschow, A. (2002) *Environmental Policy Integration: Greening Sectoral Policies in Europe*, Earthscan, London

Majchrzak, A. (1984) *Methods for Policy Research*, Sage, Newbury Park, CA, US

May, P. (1992) 'Policy learning and failure', *Journal of Public Policy*, vol 12, pp331–354

Mayring, P. (2001) 'Qualitative content analysis – Research instrument or mode of interpretation?' paper presented at the 2nd Workshop on Qualitative Research in Psychology, University of Tuebingen, Blaubeuren, Germany

Mazey, S. and Richardson, J. (1997) 'Policy framing: Interest groups and the lead up to the 1996 inter-governmental conference', *West European Politics*, vol 20, pp111–133

McQuail, D. (1992) *Media Performance: Mass Communication and the Public Interest*, Sage, Newbury Park, CA, US

Miller, R. and Blair, P. (1985) *Input–Output Analysis: Foundations and Extensions*, Prentice Hall, Upper Saddle River, US

Nilsson, M. (2005) 'Learning, frames and environmental policy integration: The case of Swedish energy policy', *Environment and Planning C: Government and Policy*, vol 23, pp207–226

Nilsson, M. and Persson, Å. (2003) 'Framework for analysing environmental policy integration', *Journal of Environmental Policy & Planning*, vol 5, pp333–359

Norberg-Bohn, V. (1999) 'Stimulating "green" technological innovation: An analysis of alternative policy mechanisms', *Policy Sciences*, vol 32, pp13–38

Pahl-Wostl, C. R., Bouwen, F., Cernesson, M., Craps, A., Dewulf, B., Enserink, N., Ferrand, P. J. G., Krykow, P., Maurel, E., Mostert, S. and Prins, C. (2003) 'Social learning in river basin management', HarmoniCOP WP2 reference document, European Commission, Brussels

Parson, E. A. and Clark, W. C. (1995) 'Sustainable development as social learning: Theoretical perspectives and practical challenges for the design of a research program', in L. H. Gunderson, C. S. Holling and S. S. Light (eds) *Barriers and Bridges to the Renewal of Ecosystems and Institutions*, Columbia University Press, New York

Pearson, P., Foxon, T., Makuch, Z. and Mata, M. (2004) ESRC Sustainable Technologies Programme project progress report (end year 1), Imperial College ESRC Sustainable Technologies Programme, London

Peters, B. G. (1998) 'Managing horizontal government: The politics of coordination', *Public Administration*, no 76, pp295–311

Peters, B. G. (2000) *Institutional Theory in Political Sciences: The New Institutionalism*, Continuum, London

Rein, M. and Schön, D. (1993) 'Reframing policy discourse', in F. Fischer and J. Forester (eds) *The Argumentative Turn in Policy Analysis and Planning*, Duke University Press, London

Rose, R. (1991) 'What is lesson-drawing?' *Journal of Public Policy*, vol 11, pp3–30

Sabatier, P. (1988) 'An advocacy coalition framework of policy change and the role of policy-oriented learning therein', *Policy Sciences*, vol 21, pp129–168

Sabatier, P. (1993) 'Policy change over a decade or more', in P. Sabatier and H. Jenkins-Smith (eds) *Policy Change and Learning: An Advocacy Coalition Approach*, Westview Press, Boulder, US

Scharpf, F. W. (1989) 'Decision rules, decision styles and policy choices', *Journal of Theoretical Politics*, vol 1, pp149–176

Schön, D. and Rein, M. (1994) *Frame Reflection: Towards the Resolution of Intractable Policy Controversies*, Basic Books, New York

Siebenhüner, B. (2004) 'Social learning and sustainability science: Which role can stakeholder participation play?', in F. Bierman, S. Campe and K. Jacob (eds) *Proceedings of the 2002 Berlin Conference on the Human Dimensions of Global Environmental Change 'Knowledge for the Sustainability Transition: The Challenge for Social Science'*, Free University, Berlin, Germany

Steen, B. (1999) *A Systematic Approach to Environmental Priority Strategies in Product Development (EPS). Version 2000 – Models and Data of the Default Method*, Chalmers University, Gothenburg, Sweden

Stemler, S. (2001) 'An overview of content analysis', *Practical Assessment, Research & Evaluation*, vol 7, no 17, available online

Suh, S. and Huppes, G. (2005) 'Methods for life cycle inventory of a product', *Journal of Cleaner Production*, vol 13, pp687–697

Teisman, G. R. (1992) *Complexe besluitvorming: Een pluricentrisch perspectief öp besluitvonning över mimtelijke investeringen [Complex Decision Making: A Pluricentric Perspective on Decisions on Infrastructure Investment]*, VUGA, 's-Gravenhage, The Netherlands

Udo de Haes, H. A., Finnveden, G., Goedkoop, M., Hauschild, M., Hertwich, E. G., Hofstetter, P., Jolliet, O., Klöpffer, W., Krewitt, W., Lindeijer, E. W., Müller-Wenk, R., Olsen, S. I., Pennington, D. W., Potting, J. and Steen, B. (2002) *Life-Cycle Impact Assessment: Striving Towards Best Practice*, SETAC-Press, Pensacola, FL, US

UN (1993) *Handbook on Integrated Environmental and Economic Accounting*, United Nations, New York

UN (1999) *Handbook of Input-Output Tables – Compilation and Analysis*, United Nations, New York

UN (2003) *Integrated Environmental and Economic Accounting – Handbook of National Accounting*, United Nations, New York

Underdal, A. (1980) 'Integrated marine policy: What? Why? How?', *Marine Policy*, vol 4, pp159–169

Underdal, A. (1989) 'The politics of science in international resource management: A summary', in S. Andresen and W. Östreng (eds) *International Resource Management: The Role of Science and Politics*, Belhaven Press, London

Weale, A., Pridham, G., Cini, M., Konstadakopolous, D., Porter, M. and Flynn, B. (2000) *Environmental Governance in Europe*, Oxford University Press, Oxford, UK

Weber, R. P. (1990) *Basic Content Analysis*, 2nd edition, Sage, Newbury Park, CA, US

Webler, T., Kastenholz, H. and Renn, O. (1995) 'Public participation in impact assessment: A social learning perspective', *Environmental Impact Assessment Review*, vol 15, pp443–463

4

Sector Environmental Analysis of Energy and Agriculture

Rebecka Engström

Introduction

In order to know how to properly integrate environmental concerns into sectoral decision making, it is important to know what the potential environmental impacts from the sector activities look like. The environment is not a single issue, but several different issues, and the impacts are caused not by one actor but many different actors. This means that different strategies must be employed for reducing different kinds of impacts, and that there might be conflicts between them. Thus the different problems may need to be ranked. Decisions taken within the sector influence environmental impacts in other sectors as well, and these may be important to assess in order to avoid sub-optimization and shifting of problems from one sector to another. Although a lot is already known of environmental problems in the two sectors studied in this book, a comprehensive assessment, determining the scale of each problem, is lacking. Thus the aims of the analysis we present in this chapter are to map as many of the environmental impacts from the sectors as possible, to evaluate the potential importance of each of them, and to discuss how the impacts could be reduced and which actors could contribute to a reduction. We will employ the term 'environmental hotspots' when we discuss the most important problems. As described in Chapter 3, most of the results come from applying the method of environmentally extended input–output analysis (IOA). Sometimes results from life-cycle assessments (LCAs) are also referred to. In a sector environmental analysis it is important to include not only the direct environmental impacts from the sector, but also all indirect effects that arise because of the activities in the sector. Direct impacts from the agricultural sector refer to those from the farms and in the energy sector to impacts from power and heat plants.

The starting point for our analysis of the agricultural sector is the total production of agricultural products in Sweden in 1999. However, by only including these, one misses a lot of important effects. The IOA embraces all kinds of input needed for this production, both in Sweden and in other countries, as well as impacts from the production of these. This means that the analysis covers, for example, production of machinery (including extraction of raw materials for such machinery), production of pesticides and fertilizers, and cultivation of seed and feed in other countries that has been imported for use on Swedish farms. Furthermore, it covers transport, processing, sale, consumption and waste handling connected to products from Swedish farms. The energy sector has a similar although slightly different definition. The energy sector analysis starts from the total production of energy carriers (electricity, heat and various fuels) in Sweden in 2000. In the case of energy carriers, much of the potential impacts arise in the consumption phase, for example when fossil fuels are combusted. Thus we find it relevant to follow impacts from not only the energy produced in Sweden, but from all that is consumed in Sweden. The analysis covers all kinds of input needed for this production, both in Sweden and in other countries, and impacts from the production of these – for example fossil fuel extraction, and production of machinery and chemicals needed for this. Moreover, it covers not only transport, processing, sale, consumption and waste handling connected to heat and power from Swedish plants, but all use of energy carriers in Sweden. The energy sector in the assessment does not contain the whole transport sector, only the transports needed for production and use of energy carriers for heating and process purposes.

In this chapter, an overview of the main problems included in the assessment will initially be given, together with a summary of how well we understood the processes and potential impacts. The level of knowledge has been proposed as an important factor determining how tractable different problems are. After that, results from the environmental assessments of each sector are presented. Each sector is introduced with some general characteristics in order to provide the reader with a basic understanding of the context, and then an exposition of resource use and emissions from the sector is given. The potential environmental hotspots are then evaluated, and this is followed by a discussion of how to reduce impacts from the sectors. Lastly, some general conclusions from the environmental sector analyses are discussed. Readers interested in more information on the analyses are referred to Engström and Wadeskog (2006) and Engström et al (2006).

Overview of Environmental Problems and Knowledge Level

This section presents commonly discussed environmental issues included in the assessments of both study sectors. The assessment is based on international and Swedish sources, and the issues are discussed from a Swedish perspective. While some general issues are described in this first part, other urgent issues that are specific for each sector are discussed later in the chapter in relation to the sectoral context.

The first issue assessed is the use of non-renewable resources. This refers to raw materials much used in society, where remaining assets do not allow unlimited use. Reduced resource use is one of the strategies used by the Swedish government to reduce environmental problems. The resources assessed in this context are mainly fossil fuels and uranium, while other resources are discussed under other headings. The magnitude of fossil fuel resources is fairly well known, although exact amounts and the possibilities of extraction are issues of discussion (see, for example, Campbell and Laherrére, 1998). Uranium resources are also quite well known (see, for example, World Nuclear Association, 2004).

Second, impacts on climate change are assessed. This refers to the accumulation of certain substances in the atmosphere which decrease heat radiation into space and so potentially contribute to increased average temperature on the earth. The generic term for these substances is greenhouse gases. According to the Intergovernmental Panel on Climate Change (IPCC), there is a high level of scientific understanding of the contributions of carbon dioxide, methane, nitrous oxide and halocarbons to radiative forcing, while the knowledge of contributions from aerosols and some other substances is low or very low (IPCC Working Group I, 2001, p8). The same report concludes that, even taking into account the remaining uncertainties, 'most of the observed warming over the last 50 years is likely to have been due to the increase in greenhouse gas concentrations' (p10). Still, changes are fairly small and mostly in line with expectations. For example, IPCC (op cit) reports that global average surface temperature increased by about 0.6°C and global average sea level rose between 0.1 and 0.2 metres in the 20th century, while the extent of snow cover has decreased by about 10 per cent since the late 1960s.

The third topic investigated is eutrophication resulting from an excess quantity of nutrients supplied to ecosystems on land or in waters. This nutrient overload can lead to unwanted changes in species composition, for example toxic algal blooms in coastal waters, and lack of oxygen in stagnant waters with large biological growth. According to a recent description of the eutrophication problem in Sweden (Miljövårdsberedningen, 2005), there are broad, unanswered scientific questions regarding nutrient flows on land, accumulation in ground and subsoil water, and the degree to which the nutrients in seabed sediments can be brought back to marine ecosystems through microbial processes and bioturbation. In particular there is insufficient knowledge of flows and ground processes of phosphorus (SOU, 2000:52, p280). Recovery of highly eutrophied lakes has been very slow since the late 1970s, although measures have been taken according to current knowledge and emissions have been reduced (ibid). None of the trends observed in surface water in the Baltic during recent decades can be tied to changed load from land (Larsson and Andersson, 2004), thus it is not known what reductions are required in order to improve the situation in the Baltic. It has been suggested that the ecosystem in the Baltic has undergone a regime shift and has now entered a eutrophic state, unlike the former oligotrophic regime, and that if such is the case, it will be very difficult to get the old ecosystem back (Miljövårdsberedningen, 2005; SOU, 2003:72).

Fourth, the use of toxic substances is explored. Toxic effects occur in humans as well as land and water ecosystems because of use and emissions of, for example, metals or hazardous chemicals. The knowledge of toxic substances occurring in society is limited (SOU, 2000:52 and 2000:53). No one knows exactly how many chemical substances there are on the market, and knowledge is lacking of many substances' health and environmental effects (SOU, 2000:53). Because of the knowledge gaps, ecologically and health hazardous substances cannot be identified, nor can risk assessments be made or adequate measures taken to limit the risks (ibid). Effects of toxic substances can in some cases be delayed for decades. Before effects are discovered, levels in the environment might be accumulated and it can be a very long time before measures to reduce the substances produce effects. There is a great lack of both knowledge and of available data in relation to this issue.

Fifth, impacts on biodiversity are assessed. Depletion of biodiversity (or biological diversity) refers to a reduction of species richness as well as of diversity of biotopes and genetic variation within species. The Millennium Ecosystem Assessment (2005), based on the work of more than 2000 authors and reviewers worldwide, established that 'changes in important components of biological diversity were more rapid in the past 50 years than at any time in human history' (p2). An important driving force is that productivity increases in agriculture require simplification of natural systems in favour of high yields. Still many things in this field are unknown. The knowledge about biodiversity's role in enhancing ecosystem resilience is incomplete (Millennium Ecosystem Assessment, 2005), and many of the negative effects associated with biodiversity depletion are slow to become apparent – a reduction in ecosystem resilience might not be visible until there is a significant change in the surrounding circumstances and the lost ability to recover is evident.

Last, air pollution is discussed, primarily referring to outdoor air quality. Air pollutants often discussed are sulphur dioxide, nitrogen oxides, carbon monoxide, volatile organic compounds, photochemical oxidants and small particles (sometimes called PM10). According to SOU (2000:52, p206), the emissions and the reductions needed to fulfil air quality objectives in densely populated areas are fairly well known.

In addition to these initial and general issues, each sector has specific issues that are discussed in relation to the sectoral context.

The Energy Sector Analysis

Table 4.1 shows some characteristics of the Swedish energy sector compared to EU-15. Sweden uses relatively little solid fuels (coal and coke), but comparatively more wood and wood wastes. The total use of energy carriers is fairly high compared to the EU-15 average. Energy carriers of Swedish origin consist mainly of biofuels and hydropower. Principally the biofuels consist of residues from production of timber and pulp and paper, including felling residues and residues later in the process (LRF and SCB, 2001). In 1997 only 0.5 per cent of the total bioenergy production came from agriculture (ibid). Imported fuels consist mainly of crude oil (Statistics Sweden, 2001b). Most of the imported oil products come from the North Sea (STEM, 2001 and 2004). A more detailed view of the Swedish use of energy carriers shows that the single largest energy carrier used in Swedish households in 1999 was electricity, followed by district heat, oil products and wood fuels (Statistics Sweden, 2001b). Among the energy carriers used in Swedish industries in the same year, electricity and bio fuels scored almost equal, followed by oil products, coke and coal (ibid). Swedish electricity was produced mainly from hydropower and nuclear, while district heat came mainly from wood fuels and waste (ibid). There were no dramatic changes in these indicators from 1999 until 2005.

Table 4.1 *Some characteristics of production and consumption of energy carriers in Sweden compared to EU-15, 1999, PJ/capita*

	Sweden	EU-15
Primary production	157	86
Gross inland consumption	241	161
Final fuel consumption – Industry	62	29
Solid fuels	*5*	*4*
Oil products	*9*	*5*
Wood and wood wastes	*21*	*1*
Final fuel consumption – Households	36	27
Solid fuels	*0*	*1*
Oil products	*5*	*6*
Wood and wood wastes	*4*	*2*

Source: Eurostat databases (Eurostat, 2006)

The sector environmental analysis included the non-renewable resources uranium and fossil fuels. About half the electricity used in Sweden comes from nuclear power, and the use of uranium for power production made up 55 per cent of the abiotic resources used in the sector. Among the fossil fuels, fuel oil was the main resource, while coal, coke, diesel and natural gas constituted only a few per cent each. The largest shares of the resources are used in power and heat production and in other countries for production of the imports, as these are the only actors using uranium. The same actors are also responsible for some of the fossil fuel use, but most of these resources are used in households and in public and non-profit institutions, together with other small users.

The most important greenhouse gas released from the production and use of energy carriers is carbon dioxide, constituting 86 per cent of the total greenhouse gas emissions from the sector according to the sector analysis. Emissions from the production of energy carriers originate from the power and heat production and from producers in other countries, from which Sweden imports energy carriers. However, even more emissions come from the use of energy, and combined emissions from many small users overshadow emissions from the large users. Household and public consumption are the main users of power and heat. About half the electricity used in Sweden comes from hydro. Hydropower reservoirs can be a source of greenhouse gas emissions, but calculations for Sweden show that water regulation has not been of any major importance in this regard as reservoirs are in locations of large differences in altitude, entailing relatively small inundated areas (Vattenfall, 1997). Brief calculations based on an LCA of Swedish hydropower showed that including these emissions in the assessment would not significantly affect the results (Vattenfall, 2005).

The sector's airborne and waterborne emissions cause eutrophication. Airborne emissions of nitrogen oxides and ammonia occur mainly in other countries as a result of imports to the sector, but also to some extent in Sweden from the use of fuels. Nitrogen oxides have the largest potential impact. Waterborne eutrophying emissions included in the IOA are nitrogen and phosphorus from biofuel cultivation in agriculture. Nutrient leakage also occurs from hydropower reservoirs; these are not included in the IOA, but brief calculations based on an LCA of Swedish hydropower (Vattenfall, 2005) showed that including these emissions would not significantly affect the results.

Toxic effects are mainly caused by the use of chemicals. In the assessment, hazard- and risk-classified chemicals are primarily used in other countries for production of fossil fuels that are then used in Sweden, although the petroleum industry in Sweden uses some as well. Fossil fuels are not included in the hazard- and risk-classified chemicals as we have used them in the IOA. However, light and water-soluble oil hydrocarbons, in particular, are so toxic that they can harm aquatic organisms at very low concentrations, for example after oil spillage from ships or from more diffuse sources (SEPA, 2005b). Steel and metal works together with households and other small actors are the largest users of fossil fuels.

Biodiversity in the energy sector is primarily connected to the use of biofuels and hydropower. Biofuels (including peat, tall oil, waste liquor and wood fuels) constitute around 40 per cent of all fuels used in stationary sources, and most of these derive from forest areas. Principally these fuels consist of residues from production of timber, pulp and paper – from both felling and later in the process. Modern forestry methods affect soil quality and biodiversity negatively in many cases. For example, a virgin forest in central Sweden has been estimated to contain around 8000 species, while a spruce plantation in the same region barely contains more than 2000 species (SEPA, 1994). It can of course be discussed how these effects should be allocated between energy and other products. Extended use of felling residues has not been found to produce any additional negative effect on soil chemistry and biological diversity, provided that certain aspects, such as a correct recirculation of ashes, are considered (Egnell et al, 1998; Jacobsson et al, 2000; Jacobsson and Gustafsson, 2001). However, long-term effects have not yet been properly investigated. Besides biofuels, hydropower also has effects on biodiversity. Hydropower dams affect nearby ecosystems both in the areas upstream that are submerged and in the downstream areas that are periodically drained (see, for example, Hjelm, 2004). Around half of the electricity in Sweden comes from hydro, and many rivers are affected by dams. Four river systems in Sweden have been protected from exploitation.

A large part of the pollutants that affect air quality emitted from the energy sector come from fossil fuel combustion. The energy sector in general is a large contributor to several of the air pollutants. Much of the emissions in Sweden come from power and heat production, as well as from use of energy by households and other small consumers. Emissions in the countries from which Sweden imports energy carriers are in most cases larger than from any single actor within Sweden. For particles smaller than 10 micrometres (PM10), the main emission source is fuel combustion for heating, especially wood fuels. Emissions from wood stoves are principally a local problem in residential districts with high densities of old wood stoves (Forsberg et al, 2005). The share of biomass in the Swedish energy system is increasing, and although there are commercially available techniques that fulfil current environmental demands, the proportion of boilers with old technology is also increasing (Hansson, 2003).

Besides these environmental aspects, the sector analysis also embraced some aspects that are more specific to the energy sector. The energy sector contributes to acidification through emissions of sulphur dioxide and nitrogen oxides. Producers of fossil fuel in other countries are large emitters of both substances, as are the power and heat producers in Sweden. On the user side, pulp and paper industries and metal works are the main contributors of sulphur dioxide. Also use of felling residues (branches and tree tops) from forests as biofuel contributes to acidification through soil processes, although this can be avoided if ashes are returned after incineration (Samuelsson and Bäcke, 1997). Radiation is another issue. About half the electricity used in Sweden comes from nuclear. During 1999, our year of study, there were no serious operational disturbances at the nuclear power plants leading to threatened security, nor any occurrence of abnormal radiation dose or doses above limit values either to staff or to the public (SKI and SSI, 2000). Treatment of nuclear waste at power plants, terminal storage of low and medium active wastes and intermediate storage of used nuclear fuel also worked well for the most part (ibid). However, the issue of terminal handling of nuclear waste has not yet been settled in Sweden and meanwhile the waste accumulates in intermediate storage. During 1999, 220 tonnes of used nuclear fuel was transported to the national intermediate storage site (SKI and SSI, 2000). Risks associated with nuclear power should not be ignored; they have not, however, been addressed in this study.

Hotspots in the energy sector

After mapping the relevant aspects for the sector, three different methods for weighting of impacts were employed in order to identify the potentially most important aspects. Table 4.2 shows the results from each of the different weighting methods. The methods use different kinds of categories, so in order to facilitate their interpretation the results are translated to the categories used throughout this chapter. However, since the three methods utilize the concepts somewhat differently, the table also shows more specifically which activity in the sector causes the hotspot in each of the methods. Based on these methods use of non-renewable resources, climate change and air quality aspects were identified as hotspots in the sector. Unfortunately, the methods used do not cover all the aspects discussed above. Thus, in order to include other aspects, separate evaluations were made by comparing the impact from the sector with Sweden's total impact. For this reason, toxicity was identified as a hotspot because of the toxicity of fossil fuels.

Table 4.2 *Hotspots in the energy sector*

Method	Hotspots	Important activities
Ecotax	Non-renewable resources Climate change	Use of uranium and fossil fuels Emissions of carbon dioxide from use of fossil fuels
EPS	Non-renewable resources Climate change	Use of uranium and fossil fuels Emissions of carbon dioxide from use of fossil fuels
Eco-indicator	Non-renewable resources	Use of fossil fuels
Share of total impact in Sweden	Air quality	Emissions of particles from fuel combustion
	Toxicity	Use of fossil fuels

Reduction of potential impacts from the energy sector

Having identified the hotspots and the sectoral activities causing these, it can be discussed how a reduced impact could be attained. A conclusion from the assessment is that many small users of fuels for heating are important target groups, along with industries such as pulp and paper and petroleum producers. Improved combustion techniques would thus help reduce impacts from the sector, and a first step is to implement the best available technology already developed. Power and heat production stand out as main contributors to many of the problems in the sector, although it has to be kept in mind that this production is, in many cases, an alternative to direct fuel use among smaller users.

Another conclusion from the assessment is that, although Sweden has a fairly low dependency on fossil fuels, their use still stands out as the most problematic in the sector. A further substitution of other fuels is therefore urgent. This could also help decrease the impacts of the production of fossil fuels in other countries for export and use in Sweden. However, when considering other fuels it is important to bear in mind the potential consequences from an increased production of these. One of the available options is to use more biofuels, both from forests and from agricultural areas. This could provide a solution for some problems in the energy sector, and in some respects also be beneficial for the agricultural sector hotspots. Yet, if not managed properly, such increased production risks severely affecting biological diversity, both within Sweden and in other countries. Using more land for production of biofuels could also increase the risk of land use conflicts with first-hand food production, as well as the production of timber, pulp, paper and other bio-materials. Alternatives to biofuels include different kinds of power production, such as nuclear, hydro and wind. At present, the Swedish line of policy is to phase out nuclear and not to further exploit rivers for hydropower. The potential negative consequences of changing this policy would also have to be handled. For nuclear, such consequences mainly involve issues of safety, concerning operation of power plants, handling and storage of nuclear waste and risk of proliferation, and in the longer run the resource base must also be considered. Increased hydropower production involves potential biodiversity depletion.

Since most kinds of energy-carrier production involve some kind of problem, an attractive line is to reduce consumption of fuels and electricity. Reduced energy use could be attained by enhancing efficiency, for example through increased insulation of houses or technological development of electrical supplies, or by changed behaviour, such as lower indoor temperatures or more careful use of electrical supplies. To attain such changes, policy coordination with other sectors is essential.

The Agricultural Sector Analysis

Table 4.3 shows some characteristics of Swedish agriculture compared to EU-15. The Swedish agricultural sector produces both animal and vegetable products: meat production is dominated by pork and beef, but among animal products dairy products are also important; vegetable products are dominated by wheat, barley and oats, used both for human consumption and as animal feed (LRF and SCB, 2001). Swedish farms have a high degree of specialization: in 1999 62 per cent of the farms were specialized in animal production, 28 per cent in crop production and only 10 per cent were mixed (counted as number of farms) (Statistics Sweden, 2001a). Compared to EU-15, Sweden has a slightly lower share of the population employed in agriculture and a substantially lower share of total area is used for agricultural purposes. Although Sweden has a degree of self-sufficiency close to 100 per cent for most animal products (numbers for 1998 in Statistics Sweden, 2001a), almost 80 per cent of manufactured feed for Swedish animals was dependent on imports in 1999 (Deutsch and Björklund, 2004). Soybean cake was the single largest important component in animal feed in the same year, and a major part of it came from Brazil (ibid). The import of soy meal to the Swedish feed industry increased fourfold during the 1990s, partly replacing domestic rapeseed meal (Mattsson et al, 2000). Sweden also exports food products, the largest export groups (in value) being grain (mostly oats exported to the US), bread and pastry, and spirits (mostly vodka) (SJV, 2003). From 1999 to 2004 slaughtering of cattle and pigs decreased.

Table 4.3 *Some characteristics of Swedish agriculture compared to EU-15, 1999*

	Sweden	EU-15
Employment in agriculture, %	3	5
Agricultural area, % of total area	7	42
Arable land, % of total agricultural area	88	64
Pasture land, % of total agricultural area	12	36
Harvested cereals, kg/capita	557	544
Harvested vegetables, kg/capita	28	140
Dairy products, kg/capita	373	314
Meat (slaughterings), kg/capita	53	72

Source: Eurostat databases (Eurostat, 2005)

The sector environmental analysis of the agriculture sector showed that non-renewable resources are mainly used in the form of fossil fuels and uranium for electricity production. Uranium makes up three quarters of the total use of non-renewable resources. The largest domestic users of both district heating and electricity connected to food purposes are private households, although the food industry also uses a lot of electricity. Around a third of the resource use occurs in other countries as a result of import and export. Fossil fuels are used in the sector in transport, in tractors and other machinery in agricultural production, and for heating purposes. About half of the fuels are used in transport and machinery and half for heating. The use of phosphorus in fertilizers shows up as only a very small part of the resource use in our analysis.

Concerning climate change, the analysis showed that almost one fifth of all greenhouse gases emitted from Swedish sources stem from the agricultural sector. In addition, almost the same amount is emitted in other countries as a result of production of products imported for use on Swedish farms and of export of products from the sector. Carbon dioxide constitutes around half of the emitted substances, mostly ascribed to import. The rest consists of nitrous oxide from processes in agricultural soils and from production of fertilizers, and methane released from digestive processes in ruminants. Emissions of greenhouse gases from the use of fossil fuels in transport and machinery were not significant compared to other sources in the sector. About 10 per cent of the carbon dioxide emissions came from households. These emissions arise mainly from gasoline for cars being driven to buy food and fossil fuel use to heat water for use in food preparation.

Eutrophication from the sector is mainly caused through leakage of nitrogen and phosphorus from fields. These emissions contribute almost half of the eutrophication from Swedish sources. Eutrophication from the sector is also caused through sewage treatment. Such pollution does not show up as significant compared to nutrient leakage from agriculture.

Toxic effects were difficult to evaluate due to lack of knowledge and data. The assessment includes release of metals into water from municipal waste-water treatment and use of risk- and hazard-classified chemicals (which include agricultural pesticides). Emissions of metals into water did not appear as a major problem in the assessment. Concerning chemicals it is interesting to note that the use of pesticides only constitutes a small part of the total use of risk- and hazard-classified chemicals in the food chain. The largest use of such chemicals in the sector derives from cement production. However, pesticides in agriculture are used in the environment with the purpose of affecting it; thus there are other risks associated with the use of pesticides which may influence both ecosystems and humans. Pesticides used abroad in the production of products imported to Sweden are treated in the IOA as if the same amounts were used in Sweden. Other studies indicate that this probably results in an underestimation of pesticide use, since a large part of the imports consists of concentrate feed from soybeans, which employs an intense use of chemicals in cropping (Cederberg and Mattsson, 2000).

Biodiversity links to the agricultural sector in many ways. In Sweden more than half of the threatened species of mammals and birds, several large groups of insects, and almost 90 per cent of the threatened vascular plants are associated with the agricultural landscape (LRF and SCB, 2001). Most of these are threatened because their habitats, such as meadows, pastureland, hedgerows and wetlands, have diminished. Eight per cent of the arable land, principally meadows and pastureland, were registered for environmental support for biodiversity in 2000 (Statistics Sweden, 2001a). In the environmental support system a further 13 per cent of the arable land was registered as organic farming land (ibid). Organic farming practices and other low-intensity arable systems have, in many cases, been found to have a higher biodiversity than conventional farms (Bengtsson et al, 2005; Cesare et al, 2003). Biodiversity depletion also occurs in other countries as an effect of import of seed and feed to Swedish farms. The cultivation of soybeans in Brazil has caused losses of biodiversity due to deforestation (Cordeiro, 2000). Land use affects biodiversity in the rest of the food chain as well. One example is infrastructure for transport. Roads affect habitats both directly, by the conversion of the original land cover into artificial surfaces, and indirectly due to fragmentation and degradation (noise, pollution, light, etc) (Geneletti, 2003). Although roads are not built exclusively for transporting agricultural products, it is a driving force contributing to road development.

The main sources of air pollution in the agricultural sector are transport and other mobile sources. In total, the transport and machinery in the food chain contribute around 10 per cent of air pollution emissions in Sweden. They also contribute emissions of the same magnitude abroad. In Sweden precursors to photochemical oxidants are mainly released from agriculture and households. Ozone is one of the photochemical oxidants and it affects vegetation by impairing growth, an effect that has been observed in crops such as wheat and potatoes. The loss in production in Swedish agriculture due to ozone is calculated to be at least a billion SEK (about 100 million euros) a year (SEPA, 2005a). It also affects human health by attacking lungs and mucous membranes.

Besides these environmental aspects, we also assessed aspects that are more specific to the sector, including soil quality. Sweden has a fairly good situation compared to many other countries, and issues that are urgent in other parts of the world, such as losses of soil organic matter or problems with soil acidity, have not been reported as severe in Sweden (Eriksson et al, 1997; LRF and SCB, 2001). However, as previously mentioned, considerable amounts of the feed used on Swedish farms comes from Brazilian soybean production, which causes serious losses of soil organic matter (Mattsson et al, 2000) and irreversible soil degradation (Cordeiro, 2000). Another aspect is that soil productivity is affected by accumulation of heavy metals such as cadmium. The largest source of cadmium is atmospheric deposition, which does not result from agricultural activities; however, cadmium is a by-product of the use of contaminated fertilizers and manure (where cadmium in imported concentrated feed causes enrichment). The imported concentrate feed brings in around 150kg/year of cadmium (LRF and SCB, 2001), although it is not known how much of this ends up in Swedish soils. Fertilizers used in Brazil for soy production have far higher cadmium content than those used in Sweden (Mattsson et al, 2000). A further issue relevant for the sector is animal health. Sweden has restrictive policies concerning the use of antibiotics and animal welfare. Since 1986, there has been a general ban on the use of antibiotics for growth-stimulating purposes. Instead husbandry, environment and disease control are supposed to compensate for the preventive effect of antibiotics. Due to Sweden's relatively isolated geographical position, favourable climate and effective control programmes, it is free from several infectious diseases that are frequent on the European continent, and the incidence of several common diseases decreased during the 1990s (LRF and SCB, 2001). Well-being of animals means more than the absence of bad health, however, and problems regarding animal transport and stress in abattoirs are discussed in the EU. In Swedish abattoirs, the number of animals that died during transport and in stables before slaughtering increased slightly in the late 1990s, perhaps due to the closure of some abattoirs, which led to longer transport (LRF and SCB, 2001). The numbers are, however, still low compared to the 1980s. Concerning genetically modified organisms (GMOs), there has been no consumer demand for such foods in Sweden, and farmers and food industries in Sweden have until recently voluntarily agreed not to use GMOs. GM crops have only been grown in field experiments. Agriculture is one of the main contributors of acidifying substances in Sweden, through emissions of ammonia from storing and handling of manure. However, the largest source of these problems in Sweden is atmospheric transport of emissions released in other countries.

Hotspots in the agricultural sector

As for the energy sector, three different methods were employed for weighting of impacts from the agricultural sector in order to identify the potentially most important aspects. Table 4.4 shows the results from each of the different weighting methods. The results are translated to the categories we use throughout this chapter and are shown together with the activities in the sector that are identified as the cause of the hotspot in each of the methods. Based on these methods, eutrophication, use of non-renewable resources and air quality aspects were identified as hotspots in the sector. Since the methods do not cover all of the aspects

Table 4.4 *Hotspots in the agricultural sector*

Method	Hotspots	Important activities
Ecotax	Eutrophication Non-renewable resources	Nutrient leakage from fields Use of uranium
EPS	Climate change Non-renewable resources	Emissions of carbon dioxide, nitrous oxide and methane Use of uranium and diesel
Eco-indicator	Non-renewable resources Air quality	Use of fossil fuels Emissions of nitrogen oxides and ammonia
Share of total impact in Sweden	Biodiversity Climate change	Farming practices Emissions of methane from cattle and nitrous oxide from soils
Closeness to ecosystem	Toxicity	Use of agrochemicals

discussed above, the impacts from the sector were also compared with Sweden's total impact. Biodiversity was then found to qualify as one of the most important issues for Swedish agriculture, since most threatened species in Sweden are connected to the agricultural landscape. Climate change also appeared important, as the sector emits around 20 per cent of all greenhouse gases from Swedish sources. For the agricultural sector another potential hotspot was evaluated based on another rationale: as discussed earlier, the analysis suffers from some important data gaps, primarily concerning waterborne emissions and chemicals, and this affected all toxicity categories in the weighting process. Because of the limited information it is hard to judge the potential importance of these aspects. However, since agrochemicals are used with the purpose of affecting the ecosystems, toxicity was identified as a potential hotspot.

Reduction of potential impacts from the agricultural sector

What are the possibilities for reducing potential impacts from the agricultural sector? Non-renewable resources are used in the agricultural sector both as fuels for transport and heating and to generate electricity used in both industries and households. A reduction of this use would involve the food consumers, who are important actors in two ways: because of their own use of electricity, heating and transport connected to the consumption of food and because they can influence the energy use in the whole sector through their choice of diet. Carlsson-Kanyama et al (2005) found that with a changed diet the energy use for production of food could be reduced by as much as 30 per cent. The most important diet changes in the study were a reduction of meat consumption and selection against vegetables cultivated in heated greenhouses, which were exchanged for legumes and root vegetables. Consumers are also important actors in other respects. The only known accessible way to substantially reduce the emissions of the greenhouse gases methane and nitrous oxide from agriculture is to reduce the number of animals and the area under cultivation (AEA, 1998a and 1998b; SJV, 2004). This can be done by importing food from other countries for Swedish consumption, which does not actually reduce the problem but merely shifts it to another part of the world. Another way is to change people's diet – for example, a diet with less animal products requires not only fewer animals, but also less acreage of arable land for growing feed for the cattle (Carlsson-Kanyama, 1998). It has also been shown that choice of diet affects nutrient leakage, since production of food for a diet with a large share of animal products results in substantially more nitrogen emitted to the air and water than for a vegetarian diet (SEPA, 1996). Thus a diet containing less animal products than at present would be beneficial for several of the hotspots in the sector.

Of course, consumers are not the only important actors in the sector. The issues of climate change, eutrophication and biodiversity are primarily connected to the farmers, and improved farming methods and techniques could be increasingly applied. Such changes include, for example, organic farming practices, techniques for handling and applying manure and fertilizers, protection zones along watercourses, and plant and animal breeding. Furthermore, farmlands could increasingly be used for cultivation of biofuels that on a larger scale can contribute to reducing emissions of greenhouse gases. Finally, actors in other countries are also important since activities in the Swedish agricultural sector cause potential impacts abroad of the same magnitude as those within Sweden. These actors might be difficult for the sector to influence; however, a first step to approach these problems is to include effects abroad when considering potential impacts from the Swedish agricultural sector.

Discussion and Conclusions

Assessments of environmental impacts from sectors are necessary in order to learn more about the performance of the sector. To know what aspects are most important to integrate in sectoral decision making it must first be known which actors and activities are causing the most important impacts. Based on such knowledge, discussions could begin on how to reduce the impacts. When assessing complex systems, an indicator such as use of energy carriers is often deployed to represent all kinds of environmental impacts. Although this might give some useful insights, many aspects are lost. The sector analyses described in this chapter give substantial information regarding different impacts, actors and activities. The assessment shows that there are substantial challenges for both sectors in the environmental area. Use of non-renewable resources, climate change, air quality aspects and toxicity are the potential hotspots for the energy sector, while eutrophication, use of non-renewable resources, air quality aspects, biodiversity and climate change seem to be the most urgent aspects for the agricultural sector. Hotspot identification provides a fuller picture of what needs special awareness in the chosen sector than an indicator would do, as it can include all the sector-specific aspects that need attention. This is crucial in order to develop measures suited to reduce the major environmental impacts from the sector.

Some environmental strategies are useful for decreasing impacts from several of the hotspots in each sector. For the two sectors accounted for here, such changes include decreased production and consumption of fossil fuels and animal products. However, in some cases a reduction of one hotspot may lead to an increase of another. Decreased use of fossil fuels might entail enhanced production of agricultural bioenergy, which might lead to other potential hotspots. Bioenergy is further examined as a policy case in Chapters 5 and 6. Our first conclusion is that it is important to remember that the environment is not one issue, but several, and a comprehensive view of different kinds of impacts is needed in order to find solutions that actually contribute to decreased impacts, rather than decreasing one hotspot while increasing another type of impact.

Although sector-specific angles have to be taken into consideration, there are also similarities between the sectors. Climate change is an issue that is a hotspot in both sectors. Figures 4.1 and 4.2 show greenhouse gas emissions distributed among different actors in the two sectors. The figures show that in both sectors a lot of the emissions occur in other countries, and that also within Sweden considerable amounts of greenhouse gases are emitted by actors outside the sectors as they are commonly defined in the policy system. These findings are valid not only for climate change, but also for other hotspots. Moreover, it is likely that this

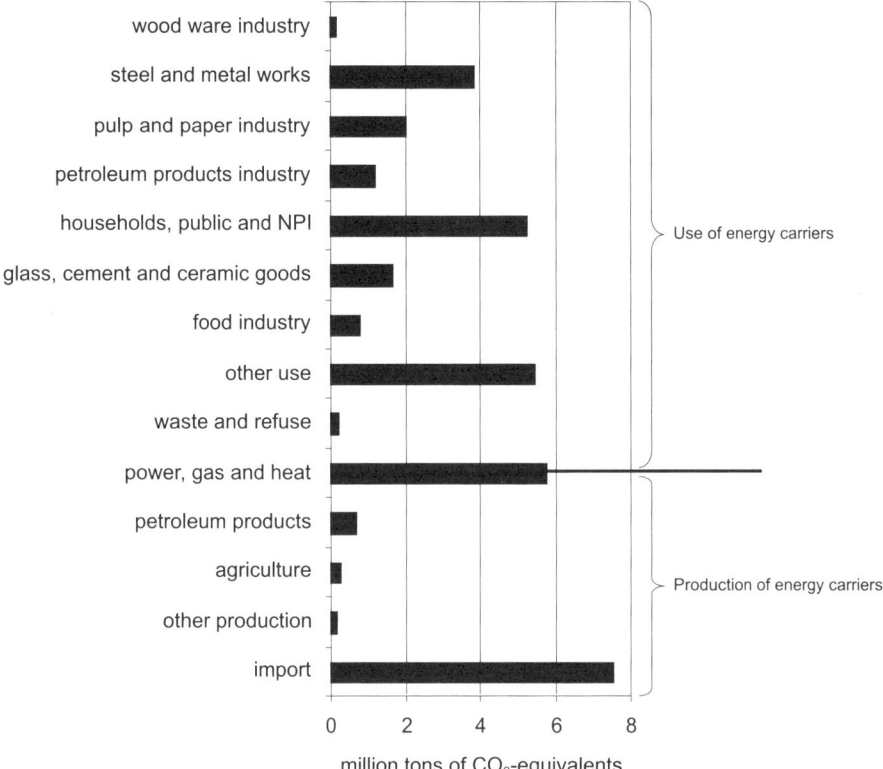

Note: NPI = non-profit institutions

Figure 4.1 *Emissions of greenhouse gases from the energy sector*

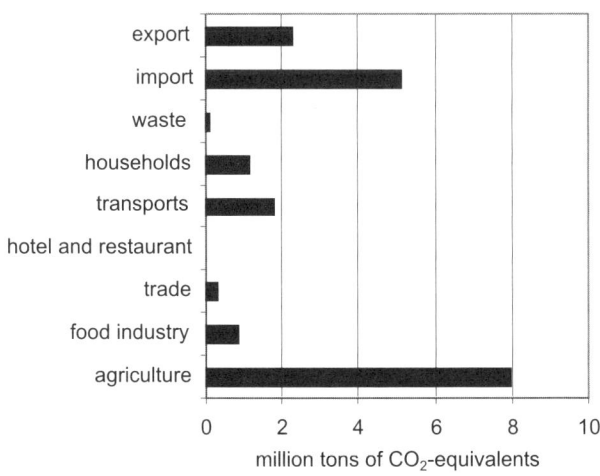

Figure 4.2 *Emissions of greenhouse gases from the agricultural sector*

would be found not only for the two sectors shown here, but also for other sectors. Finnveden et al (2002) investigated indirect environmental influence from the Swedish defence sector, and found the indirect effects to be as large as the direct impacts from the sector. Our second conclusion is, therefore, that indirect effects are important to include in assessments of environmental impacts from sectors.

In this chapter we have identified hotspots in the energy and agricultural sectors. The hotspot evaluation separates big issues from small ones and points out the most urgent questions to handle for each sector. But are these issues also the issues that receive the most attention in policies? The following chapters will explore environmental policy integration in the sectors. Although we do not establish a direct causal link between EPI and environmental impacts, the later chapters can be read with the hotspots from this chapter in mind. Is policy attention related to the issues' relative importance? If not, what decides the integration of environmental issues into policymaking?

References

AEA (1998a) *Options to Reduce Methane Emissions*, AEAT-3773: Issue 3, European Commission, Brussels

AEA (1998b) *Options to Reduce Nitrous Oxide Emissions*, AEAT-4180: Issue 3, European Commission, Brussels

Bengtsson, J., Ahnström, J. and Weibull, A.-C. (2005) 'The effects of organic agriculture on biodiversity and abundance: A meta-analysis', *Journal of Applied Ecology*, vol 42, pp261–269

Campbell, C. J. and Laherrére, J. H. (1998) 'The end of cheap oil', *Scientific American*, March, pp78–83

Carlsson-Kanyama, A. (1998) 'Climate change and dietary choices: How can emissions of greenhouse gases from food consumption be reduced?', *Food Policy*, vol 23, pp277–293

Carlsson-Kanyama, A., Engström, R. and Kok, R. (2005) 'Indirect and direct energy requirements of city households in Sweden: Options for reduction, lessons from modelling', *Journal of Industrial Ecology*, vol 9, pp221–236

Cederberg, C. and Mattsson, B. (2000) 'Life cycle assessment of milk production – A comparison of conventional and organic farming', *Journal of Cleaner Production*, vol 8, pp49–60

Cesare, P., Wossink, A., Giesen, G., Vazzana, C. and Huirne, R. (2003) 'Evaluation of sustainability of organic, integrated and conventional farming systems: A farm and field-scale analysis', *Agriculture, Ecosystems and Environment*, vol 95, pp273–288

Cordeiro, A. (2000) *Sustainable Agriculture in the Global Age. Lessons from Brazilian Agriculture*, Naturskyddsföreningen, Stockholm

Deutsch, L. and Björklund, J. (2004) 'Made in Sweden? Re-defining the Swedish animal production system', in L. Deutsch 'Global trade, food production and ecosystem support: Making the interactions visible', doctoral thesis, Stockholm University, Stockholm

Egnell, G., Nohrstedt, H.-Ö., Weslien, J., Westling, O. and Örlander, G. (1998) *Miljökonsekvensbeskrivning (MKB) av skogsbränsleuttag, asktillförsel och övrig näringskompensation* [*EIA of Wood Fuel Harvesting, Ash Application and other Nutrient Compensation*], Skogsstyrelsen, Jönköping, Sweden

Engström, R. and Wadeskog, A. (2006) 'Environmental impact from a sector: Production and consumption of energy carriers in Sweden', *Progress in Industrial Ecology*, submitted manuscript

Engström, R., Wadeskog, A., and Finnveden, G. (2006) 'Environmental assessment of Swedish agriculture', *Ecological Economics*, available online

Eriksson, J., Andersson, A. and Andersson, R. (1997) *Tillståndet i svensk åkermark* [*State of Swedish Arable Soils*], Naturvårdsverket, Stockholm

Eurostat (2005) http://epp.eurostat.cec.eu.int/, Agriculture and Fisheries, data accessed 14 November 2005

Eurostat (2006) http://epp.eurostat.cec.eu.int/, Environment and Energy, data accessed 6 February 2006

Finnveden, G., Wadeskog, A., Eriksson, B. N., Johansson, J., Palm, V., Åkerman, J. and Hedberg, L. (2002) *Indirekt miljöpåverkan från försvarssektorn* [*Indirect Environmental Effects from the Defence Sector*], Totalförsvarets Forskningsinstitut, Stockholm

Forsberg, B., Hansson, H. C., Johansson, C., Areskoug, H., Persson, K. and Järvholm, B. (2005) 'Comparative health impact assessment of local and regional particulate air pollutants in Scandinavia', *Ambio*, vol 34, pp11–19

Geneletti, D. (2003) 'Biodiversity impact assessment of roads: An approach based on ecosystem rarity', *Environmental Impact Assessment Review*, vol 23, pp343–365

Hansson, H. C. (2003) 'Biobränsle – Hälsa – Miljö. Preliminär slutrapport, Sammanfattning' ['Biofuel – Health – Environment. Preliminary final report, summary'], ITM, Stockholm

Hjelm, S. (2004) *Vattenkraft – Miljöpåverkan och åtgärder* [*Hydro Power – Environmental Impacts and Measures*], Energimyndigheten och Elforsk, Stockholm

IPCC Working Group I (2001) 'Summary for policymakers', report of Working Group I of the Intergovernmental Panel on Climate Change, UNEP, Geneva, Switzerland

Jacobsson, S. and Gustafsson, L. (2001) 'Effects on ground vegetation of the application of wood ash to a Swedish Scots pine stand', *Basic and Applied Ecology*, vol 2, pp233–241

Jacobsson, S., Kukkola, M., Mälkönen, E. and Tveite, B. (2000) 'Impact of whole-tree harvesting and compensatory fertilization on growth of coniferous thinning stands', *Forest Ecology and Management*, vol 129, pp41–51

Larsson, U. and Andersson, L. (2004) *Varför fosfor ökar och kväve minskar i egentliga Östersjöns ytvatten* [*Why Phosphorus Increases and Nitrogen Decreases in Surface Water in the Baltic Sea*], Stockholm University, Stockholm

LRF and SCB (2001) 'Miljöredovisning för svenskt jordbruk 2000' ['Environmental Reporting for Swedish Agriculture 2000'], Lantbrukarnas Riksförbund, LRF, Stockholm

Mattsson, B., Cederberg, C. and Blix, L. (2000) 'Agricultural land use in life cycle assessment (LCA): Case studies of three vegetable oil crops', *Journal of Cleaner Production*, vol 8, pp283–292

Miljövårdsberedningen (2005) 'Strategi för hav och kust utan övergödning' ['Strategy for sea and coast without eutrophication'], Promemoria 2005:1, Regeringskansliet, Stockholm

Millennium Ecosystem Assessment (2005) *Ecosystems and Human Well-Being: Biodiversity Synthesis*, World Resources Institute, Washington, DC

Samuelsson, H. and Bäcke, J.-O. (1997) *Effekter av skogsbränsleuttag och askåterföring – En litteraturstudie* [*Effects from Wood Fuel Harvesting and Ash Application – A Literature Study*], Skogsstyrelsen, Jönköping, Sweden

SEPA (1994) *Biological Diversity in Sweden: A Country Study*, Naturvårdsverket, Stockholm

SEPA (1996) *Biff och bil? Om hushållens miljöval* [*Beef and a Car? About Environmental Choices of Households*], Naturvårdsverke, Stockholm

SEPA (2005a) 'Ozon' ['Ozone'], page linked from www.naturvardsverket.se homepage via Föroreningar page, accessed 12 April 2005

SEPA (2005b) 'Olja och avfall i havet' ['Oil and waste in the ocean'], page linked from www.naturvardsverket.se homepage via Föroreningar page, accessed 3 June 2005

SJV (2003) 'Sveriges livsmedelsexport 2002' ['Export of food from Sweden 2002'], Jordbruksverket, Jönköping, Sweden

SJV (2004) 'Förutsättningar för en minskning av växthusgasutsläppen från jordbruket' ['Requirements to reduce greenhouse gas emissions from agriculture'], Jordbruksverket, Jönköping, Sweden

SKI and SSI (2000) *Säkerhets- och strålskyddsläget vid de svenska kärnkraftverken 1999* [*The Situation of Safety and Protection Against Radiation 1999*], Swedish Radiation Protection Institute and Swedish Nuclear Power Inspectorate, Stockholm

SOU (2000:52) 'Framtidens miljö – Allas vårt ansvar!' ['Environment of the future – All our responsibility'], Government Committee Report, Regeringskansliet, Stockholm

SOU (2000:53) 'Non-hazardous products: Proposals for implementation of new guidelines on chemicals policy', Government Committee Report, Regeringskansliet, Stockholm

SOU (2003:72) The sea – Time for a new strategy', Government Committee Report, Regeringskansliet, Stockholm

Statistics Sweden (2001a) *Jordbruksstatistisk årsbok 2001* [*Yearbook of Agricultural Statistics 2001*], Statistics Sweden, Örebro, Sweden

Statistics Sweden (2001b) 'Årliga energibalanser 1998–1999' ['Yearly energy balance sheets 1998–1999'], Statistics Sweden, Stockholm

STEM (2001) 'Energiläget i siffror' ['Energy situation in figures'], Energimyndigheten, Eskilstuna, Sweden

STEM (2004) 'Energiläget i siffror' ['Energy situation in figures'], Energimyndigheten, Eskilstuna, Sweden

Vattenfall (1997) *Vattenfalls livscykelanalyser av elproduktionen* [*Vattenfall's Life Cycle Studies of Electricity*], Vattenfall, Stockholm

Vattenfall (2005) *Vattenfall AB Elproduktion Nordens Certifierade Miljövarudeklaration EPD för el från Vattenfalls vattenkraft i Norden* [*EPD for Electricity from Vattenfall's Hydro Power in the Nordic Countries*], Vattenfall, Stockholm

World Nuclear Association (2004) www.world-nuclear.org/info/inf75.htm, accessed 1 July 2005

5
Policy Framing and EPI
in Energy and Agriculture

Måns Nilsson, Katarina Eckerberg, Lovisa Hagberg,
Åsa Gerger Swartling and Charlotta Söderberg

As discussed in Chapter 3, our analysis of environmental policy integration (EPI) into sectors combines several approaches. Chapter 4 examined the environmental outcomes of sector activities through a sector environmental analysis. This is a useful departure point for an EPI study as it helps us to characterize the problem. However, through the learning approach taken in this volume, EPI is not measured through impact analysis but is concerned with how policymakers incorporate aspects of environmental sustainability into their frames of reference through conceptual learning processes. The purpose of the present chapter is to analyse this policy framing and to what extent (and in what ways) environmental sustainability issues have become part thereof, based on the methodology outlined in Chapter 3. We examine the evolution of energy, agricultural and bioenergy policy in Sweden by way of three discussions. First, we present a content analysis of major policy bills to trace the emergence of the different environmental issues in government bills, covering the period from 1997 to 2004. Second, we identify the policy frames that coexist in each sector and their main features, including their relation to environmental policy, and which actors and organizations adhere to the different frames. And third, we discuss their evolution over time by way of learning and the ways in which they have shaped policy and policy preferences in different directions. The analysis addresses both the policy itself and the underlying argumentation and reasoning, covering the period from the late 1980s to the mid 2000s. But first, a broader contextual background will set the scene for the analysis.

Contextual Background

The end of the 20th century was a time of rapid political and societal change for the world, including Europe and Sweden. Contextual changes during this time, both internationally and in Sweden, form a necessary backdrop against which the policy development may be understood, and these should be kept in mind as we go into our case studies.

The end of the cold war period and the remarkably speedy and relatively peaceful dismantling of the Soviet empire had two major implications. First, Swedish neutrality policy lost its significance.[1] It was no longer as critically important to safeguard independent and resilient national self-sufficiency in energy and food supply in case of a war between the superpowers. After many years of scepticism towards European cooperation, the Social Democrats suddenly changed direction in the early 1990s, promoting Swedish membership and liaising with the Liberals and Conservatives. After a referendum in 1994, Sweden joined the EU in 1995. Second, the Western-style capitalist democracy was basically left unchallenged as the preferred development model. Overarching liberal economic policy ideas swept across the world, anchored in Thatcher's and Reagan's economic policies of the 1980s. In Europe, political and market integration went forward at full speed. European leaders created the EU through the Maastricht (1992) and Amsterdam Treaties (1997). In their vision, the European project was built on the idea of a 'single market' and the free movement of capital, labour, goods and services. Deregulation of markets such as energy, telecommunications and transport was high on the agenda. Liberal winds also swept through Swedish society, including the government offices, leading to deregulation of, for instance, the financial markets in the late 1980s and the agricultural markets in 1990.

During the same period, environmentalism reached the political mainstream. In the 1970s the international community gradually became more attentive to environmental degradation caused by human activities worldwide. In Sweden, which was one of the pioneering countries in the environmental arena and the host country of the 1972 UN Conference on Human Development, a growing number of policy actors became increasingly concerned with environmental impacts. The emerging political debate intensified throughout the 1980s. This decade was also when the climate change issue became a political priority both internationally and in Sweden. The threats of global warming had been initially discussed in the 1970s within segments of the scientific community and continued to rise in interest within academia. By the mid 1980s the national debate had gained momentum and reached a point where actors across a range of both governmental and non-governmental institutions recognized the need to put climate change on the national policy agenda. The political debate rested largely upon scientific claims.

The 1980s also saw an explosion in public environmental consciousness in Sweden, as elsewhere, fuelled by media attention to dying forests, the Chernobyl nuclear accident, chemical accidents in the Rhine, climate change and acidification, and, more locally, massive seal deaths along the coasts of Scandinavia. At the time, the public ranked the environment among the top priorities for political action – a position it had not been close to before and has since lost (Holmberg and Weibull, 2006). In short, a political momentum was created that for the first time allowed environmental values to be prominently treated in governmental processes at local, national and international levels. At the global level, UN conferences and commissions, such as WCED in 1987 and UNCED in 1992, gave the issue international political flare and status (see Chapter 1). At the national level, public awareness led to immediate repercussions in the political arena in

Sweden, as in many other countries, and provided the necessary leverage for the Green Party to grow and enter the Parliament in 1988. Public administration was also affected; already in the 1980s the government took on an explicit strategy to include environmentalists within its ranks. Many leading 'greens' developed political careers or were employed as civil servants.

Since 1990, however, there has been a tendency of decreasing attention to environmental issues in domestic policy. Drawing on a quantitative content analysis of the presence of an environmental agenda in the governmental policy declarations and debates across political parties represented in parliament, a recent report from the Swedish Society for Nature Conservation (SSNC), Sweden's largest environmental NGO, finds that ever since the peak years 1988 and 1990, the attention paid to environmental considerations has decreased, and this substantially over the period 1991–1993. Although the attention to environmental issues increased somewhat in 1994 and 1995, the general trend has been reduced attention, with 2001 as the lowest score since 1993 (SSNC, 2003, p31). In the following section, we present a more recent content analysis regarding to what extent environmental concepts have shown up in sectoral policy bills.

Content Analysis of Energy and Agricultural Bills

A content analysis was made to trace the emergence of different environmental issues in governmental bills (Engström et al, 2006). The analysis uses the same categories as the sector analysis in Chapter 4. For the energy sector, data is based on the major energy bills of 1998 (Prop, 1996/97:84) and 2002 (Prop, 2001/02:143) and the energy chapter in the annual budget bill for the last nine years. For the agricultural sector, data is based on the agriculture chapter in the annual budget bill for the last nine years as well as the 'Sustainable agriculture and fisheries' bill (Prop, 1997/98:2) and 'The environment and rural development programme 2000–2006' (Jordbruksdepartementet, 2000). Regarding the contents analysis of the annual budget, it should be noted that the volume of text has increased significantly from 21 pages in 1997 to 98 pages in 2005 for energy and, less dramatically, from 67 to 124 for agriculture. Increases in contents of environmental themes must be put in relation to this overall increase.

The content analysis suggests that there is a relatively narrow environmental focus in both energy and agriculture, although in agriculture it is somewhat broader. It appears that the sectoral policy system has difficulties handling more than two environmental issues. Second, there seems to be considerable path dependency in the topics addressed, with basically the same two issues throughout the studied time period. In energy, there is a persistent focus on climate change and on renewable resource use, and an overall negligence of any other issues, until 2002, when issues such as biological diversity, air quality and acidification begin to be discussed. We see the same patterns in the large energy bills in 1997 and 2002 (see Figure 5.1). In agriculture, there are relatively few changes, with an overall focus on toxicity (the use of agrochemicals) and biological diversity throughout the time period. Acidification appears to have been removed from the agenda. Also in the major agricultural bills of 1997 and 2000 there is a complete negligence of climate change. Not until 2005 does it emerge at all (see Figure 5.2).

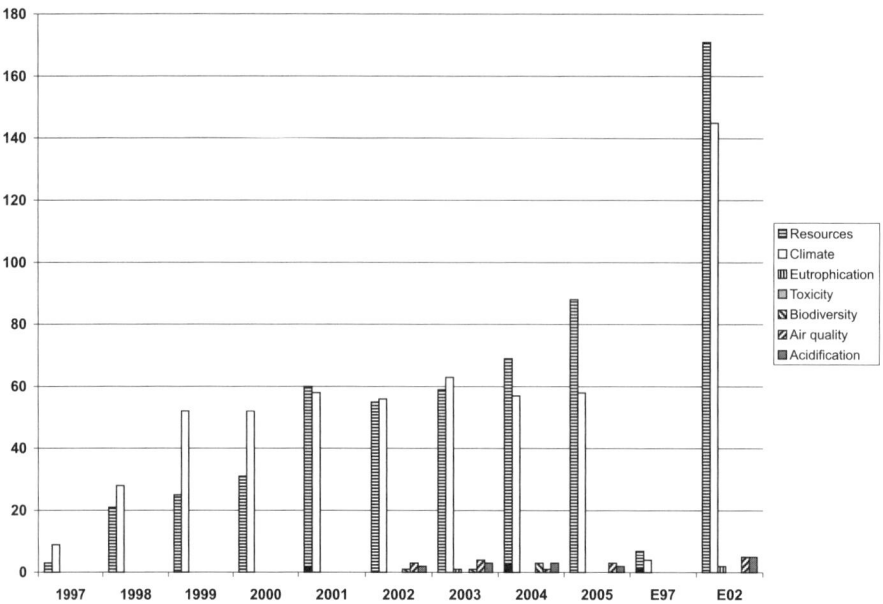

Note: Vertical axis shows number of instances of categories being mentioned in the documents. For the energy sector, data is based on the major energy bills of 1997 (**E97**) (Prop. 1996/97:84) and 2002 (**E02**) (Prop. 2001/02:143) and the energy chapter in the annual budget bill for the last nine years.

Figure 5.1 *Content analysis of energy bills and section of budget bills, 1997–2005*

The content analysis merely counts the number of mentions of different environmental categories in documents. Therefore, the interpretation of the results does not allow any direct conclusions with regard to EPI to be drawn. The content analysis sets the scene for the frames analysis rather than representing a critical part of it. However, to the extent to which the content analysis mirrors the findings in the following framing analysis, it can strengthen our conclusions.

Framing of Energy Policy

Energy policy has been intertwined with environmental protection concerns ever since the early 20th century. For instance, much of the nature conservation movement was ignited by the hydropower developments that Sweden pursued to ensure national independence and secure industrial supply. Hydropower became the major issue of environmental political conflict, as massive developments of hydropower in the north of Sweden took place from the 1910s to the 1960s (Vedung and Brandel, 2001). In the early 1960s, a broad political agreement was reached to protect remaining virgin rivers from exploitation. At the same time, dependency on oil imports had increased rapidly during the economic growth period of the 1950s and 1960s (Bergman, 2001). In order to decrease oil dependence, the expansion of nuclear power was given a high political priority. Environmentalists had promoted nuclear power earlier on, as an alternative to hydropower

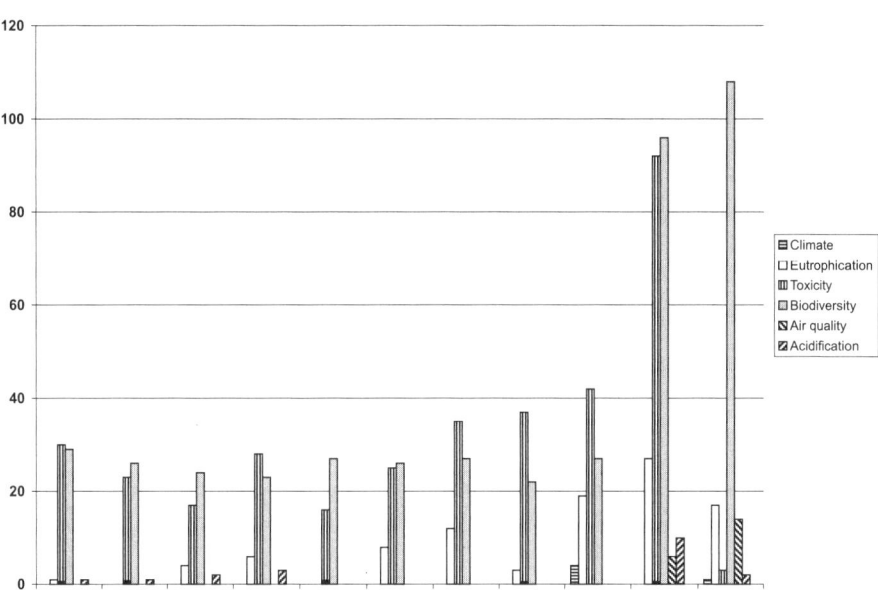

Note: Vertical axis shows number of instances of categories being mentioned in the documents. For the agricultural sector, data is based on the bill 'Sustainable agriculture and fisheries' (Prop. 1997/98:2) (**J97**) and 'The environment and rural development programme 2000–2006' (Jordbruksdepartementet, 2000) (**J00**), as well as the agriculture chapter in the annual burget bill for the last nine years.

Figure 5.2 *Content analysis of agricultural bills
and section of budget bills, 1997–2005*

expansion, but, inspired by public concern, new countermovements emerged (Anshelm, 2000). Policymakers faced a difficult equation of nuclear resistance set against higher prices on imported fuels and the newly established decision not to develop more of the undeveloped rivers for hydropower. Since then, energy has been at the centre of Swedish politics, igniting many hopeful attempts to create lasting political agreements. The first major political energy policy agreement was made in 1975 (Lundgren, 1978; Vedung, 1982). In 1976 the nuclear issue was powerful enough to have a decisive influence on the governmental elections. Capturing a growing public nuclear resistance, the Centre Party grew dramatically in size and managed in 1976 to remove the Social Democrats (who had enjoyed uninterrupted governmental rule since 1932!) from government. In 1980, after the Three Mile Island incident in the US, the Social Democrats' leader Olof Palme realized that the nuclear issue was such a political minefield that it risked breaking up his party, and he pushed for a referendum. The outcome of this referendum forms an important backdrop to the discussion because its interpretation is still an issue of debate: voters were given not two but three options to choose from, and the somewhat ambiguous in-between option, with the recommendation to phase out nuclear power eventually, but only as far as deemed possible considering future energy demand and potential implications on employment and welfare, attracted the most votes.

Three energy policy frames

Up to the 1980s, the dominant policy frame was 'energy-as-infrastructure'. Core concerns of this frame were closely associated with classic Social Democrat priorities: economic welfare, industrial production and jobs. The overriding objective of energy policy was to safeguard Sweden's industrial interests, which meant reliable supply and low prices. Energy-as-infrastructure held a central-planning perspective on the sector and preferred political intervention and regulatory measures over market solutions or economic instruments in all policy domains. Environmental concerns were completely absent at the time, as the then Minister for Energy testifies:

> *The energy function in government was supply of energy – period. That was the deal. This is what created conflicts with among others Vattenfall (the state energy producer), who was my worst opponent [....] it was an agency in those days and they could not understand how an energy minister could put any environmental aspects whatsoever on energy production. [...] This perspective is what we need to understand to be able to make understandable what has happened in the last 20 years.* (Interview, Olof Johansson, 2004)

Energy-as-infrastructure is, however, not necessarily incompatible with environmental concerns. It can in principle take environmental values into account. However, as we will see later on, it shapes the ways in which environmental concerns are addressed.

The dominance of energy-as-infrastructure was coupled with the dominance of a policy network, an 'iron triangle', between industry, the labour unions and the government. At the centre was a strong industrial user and producer integration (in the industrial meaning of the word) that paved the way for nuclear expansion, a 'development block' of industrial energy users, equipment producers and energy producers (Kaijser, 2001). The coalition restricted outside access to the policy process and maintained energy as basically a non-political issue. In the 1980s, however, the development block began to disperse. One notable reason was that the equipment producer ASEA (arguably its main hub) merged with a Swiss company to form the Swiss-based ABB, lost its dependency on the Swedish market and became just another external actor (Kaijser, 2001). This dissolution of power and influence paved the way for other actors and frames to enter the energy policy arena. Still, energy-as-infrastructure lives on to this day. It has a political support basis within the Social Democrats with support, albeit arguably dwindling, in the old bureaucracies and power industries.

In the 1970s larger societal streams of change disturbed the established political order. Challenging established orders and authorities became a norm in many parts of society, and the environmental issue, as well as nuclear resistance, became a powerful political tool. In the energy sector, a new generation of energy systems analysts began to challenge the established wisdoms and mainstream planning ideas. They argued that a transition to a renewable energy system was not only preferable to a system based on uranium and fossil fuels, but was also highly achievable, and that it was not really a technical challenge but rather a social and institutional one. These ideas, coupled with the growing environmentalism, formed the basis of a new 'energy-as-risk' frame. This frame became important in

Swedish policy during the time of the nuclear referendum in 1980s, and later on it became a driving force behind the policy positions taken in the late 1980s (see below). The core of energy-as-risk is based on notions about society's relationship to nature, which tend to see the environment as a restriction on economic activities and growth. The frame has a policy core of decentralization, conservation and resource efficiency. Thus environmental values are not only compatible with the frame but are inherent to its fundamental premises (Nilsson, 2005). Energy-as-risk has a political basis among the Greens, the Centre, and parts of the Socialists and the Social Democrats (both parties are divided on environmental issues), with support from environmental academicians and green interests groups.

In the 1980s and 1990s the dominant view of the state as key provider of development was dismantled throughout society. As a result, in the early to mid 1990s, the energy sector saw increasing waves of market thinking. The governmental agency and power producer was turned into a corporation, Vattenfall. It no longer had any responsibility for implementing policy, only generating profit to its shareholder (the State). Vattenfall, and other Swedish producers, became internationally oriented and less dependent on the Swedish market. These policy changes and supporting discourses formed an 'energy-as-market' frame, essentially a neo-liberal paradigm. Energy-as-market emphasizes the market's ability to adjust and create the best conditions for developing wealth and opportunities. Markets are the preferred instrument for resource allocation and for expressing environmental, social or economic values; economic instruments are the preferred means of governmental intervention. In principle environmental concerns can be made compatible, if expressed in market terms, but in practice the overriding premise of policy is liberalization with as few 'disturbances' as possible, such as policy interventions to correct market failures. Energy-as-market has a political basis among Conservatives, Liberals and, to some extent, Social Democrats, with support in the bureaucracies of the Ministries of Finance and Industry and among industry associations and academic economists. Table 5.1 presents some summary features of the three energy policy frames.

Table 5.1 *Summary of three energy policy frames*

	Energy-as-risk	Energy-as-infrastructure	Energy-as-market
Underlying problems	Environmental and health risks of energy production	Inadequate supply	Inefficient market systems
Policy preferences	Central planning and regulation steering towards absolute ecological targets	Public investment and planning, central interventions	Market instruments, minimal public interventions
Core values and goals	Conservation, resource efficiency, environmental protection	Welfare state-building, employment, welfare creation, national efforts	Democratic capitalism, efficiency, welfare creation, international efforts
Scale view	National, long term	National, long term	International, short term

Source: adapted from Nilsson (2005)

The three frames have at different times all profoundly influenced the formation of energy policy. They have also evolved and reframed over time as a result of conceptual learning across frames. This has enabled policy change and EPI. In the following sections, three policy rounds will be examined in more detail.

Around the energy policy decision in 1988

In the 1980s oil prices and electricity prices fell and energy lost its high rank on the political agenda. No decision was made regarding when to start implementing the referendum result and close the first nuclear reactor. Little effort was made to facilitate phase-out through energy efficiency and renewable energy technology development (Kaijser, 2001). However, late in the decade a new agenda on environmental and energy issues quickly took shape, no doubt catalysed by the Chernobyl meltdown in April 1986. The new Minister for Energy and Environment, Birgitta Dahl, pushed a radical policy agenda in the bill 'Energy policy for the 1990s' in 1988. Following this bill the parliament decided that two nuclear reactors were to be shut down in 1995 and 1996. At this time the climate change issue was also adopted as a political priority. It was decided that Sweden should not increase its carbon dioxide emissions over time. Furthermore, the protection of rivers from hydropower exploitation was restated. Environmental concern was at the centre of the energy policy agenda, learning from and implementing ideas and concerns within energy-as-risk.

However, the phase-out agenda stirred up such political tension that a reversal in policy was finally inevitable. A committee of inquiry, EL90, was appointed in December 1989 to investigate the implications for industrial competitiveness of nuclear phase-out (SOU, 1990:21). (It is interesting to note that this investigation was made *after* the decision had been made, see also Vedung and Brandel, 2001.) EL90 concluded that the nuclear phase-out, the river protection and the carbon dioxide reduction objectives had to be revoked in order to secure industrial competitiveness. With the report in hand, the government could claim that, although they wanted to start phasing out nuclear power, their hands were tied. Indeed, the energy policy decision in 1988 had stirred up massive protests from many camps: within the Social Democrats, the labour unions, the energy producers and heavy industry. Despite Dahl's strong position in the party and the government, the Prime Minister, Ingvar Carlsson, moved energy issues back to the Ministry for Industry in 1990. In 1994 Ms Dahl left politics and was appointed Speaker of the Parliament, a formal 'master of ceremonies' position without an active political role. The retreat from the 1988 energy policy decision and its nuclear phase-out agenda was completed with the 1991 energy agreement and the ensuing bill 'On energy policy' (Prop, 1990/91:88). It was decided to postpone the closure of the first nuclear reactor until 1999. The agreement established that phase-out was contingent on energy conservation measures, competitive prices and a supply of alternative technologies. This agreement replaced the use of negotiated timetables with an institutional arrangement consisting of annual evaluations on whether conditions for commencing the phase-out were met.

This round initially displays a strong reframing and an almost wholesale adoption of energy-as-risk. Ms Dahl advanced the risk frame forcefully, not only in energy policy, but also forcing, for example, the paper and pulp industry to phase out the use of chlorine for bleaching. In this way, a green agenda was established. But the subsequent backlash suggests Ms Dahl might have gone too far and alienated the political mainstream and sectoral interests. Therefore there was little enduring reframing and EPI. However, we can now see that the agenda ignited conceptual learning in parts of the government and was a seed of change for further EPI to be realized later on. Energy-as-infrastructure had been challenged and was to be further challenged by the winds of liberalization and energy-as-market. This period is dominated by national party politics and, since it precedes Swedish membership of the EU, there was little attention paid to the EU policy agenda in national energy policy efforts.

Around the energy agreement in 1997

In the early 1990s it was widely agreed that the political parties had to reach a broad (across the left–right coalitions) political agreement on energy policy. Sweden needed a stable and predictable policy that would not change just because there was a shift in government. In the summer of 1994, the government set up a committee of inquiry to prepare for such an agreement. This 'Energy Commission' was set up as a truly ambitious policy-learning enterprise. It was well resourced to take stock of, and learn lessons from, facts and experience; it commissioned more than 60 sub-reports, involved over 100 experts and published its findings in five thick volumes (SOU, 1995:139–140). Its chairperson orchestrated a political agreement between the Social Democrats (who had regained government power in 1994) and two smaller, moderately pro-nuclear parties (the Liberals and the Christian Democrats), which would be sufficient for a parliamentary majority.

However, an unexpected turn of events occurred, and whatever learning took place in the committee, it failed to reach the policy decision stage. As we saw in Chapter 1, at this point the new Prime Minister, Göran Persson, launched a new 'green people's home' agenda for Sweden. This agenda came to have strong implications for the energy sector. In early 1997 the government announced that a new agreement on energy policy had been reached with the Socialists and the Centre, neither of which had been part of the consensus in the Energy Commission. The agreement included quickly closing the nuclear reactors in Barsebäck, the first before 1 July 1998 and the second before 1 July 2001 (Prop, 1996/97:84). Barsebäck 1 was in fact closed in November 1999. The closure of the second reactor was postponed several times in the early 2000s because of concerns about supply safety and environmental impacts, but after a renewed agreement between the Social Democrats, the Left and the Centre, Barsebäck 2 closed in June 2005.

The policy at this time is strongly framed by both energy-as-risk and the old energy-as-infrastructure. The means of becoming greener was definitely not through markets or people's choices but through traditional Social Democratic instruments: public investments and large-scale planning. It was as if the 1988 decision had been reshaped and tempered with a strong ecological modernization discourse that emphasized economic growth as consistent with environmental protection (Hajer, 1995). With reference to a glorious past of being a forerunner in welfare policy, the Social Democrat leaders sketched a prosperous future of being a forerunner in environmentalism, achieved through bold governmental planning, policy and public investment.

Around the energy agreement in 2002

In the aftermath of the Barsebäck 1 closure, the energy-political collaboration between the Social Democrats, the Centre and the Socialists continued. In 2002 they presented a new energy policy bill that confirmed the overarching goals and aspirations of energy policy, but also changed certain things (Prop, 2001/02:143). First, it specified a new model for deciding on the closure of nuclear reactors; inspired by a German attempt, it was now a matter of a negotiation between the government and the nuclear power industry. An agreement in accordance with this model would be based on an agreed maximum energy production in the remaining reactors. This volume could then be distributed freely over time. Second, voluntary agreements were being developed on issues such as energy efficiency (Ds, 2001:60). Third, the use of economic instruments, such as subsidies for renewable energy, was de-emphasized. Fourth, research and development schemes were also de-emphasized, compared with the programmes in the late 1990s. Fifth, it recognized that conventional taxation instruments performed poorly in the new integrated international electricity market. For instance, the national tax on fossil combined heat and power (CHP) production had very poor effects; instead of inducing a shift towards other fuels in Sweden and curbing domestic fossil-based power production, it shifted fossil-based production abroad. The Swedish CHP plants were standing idle and distributors instead purchased the marginal power from Denmark's coal-fired plants.

The new policy agenda reflected a new market context in which the government could not devise an 'energy plan' as in the past (SOU, 2003:80). The belief that publicly funded research and development would lead to the commercialization of new technologies weakened substantially (Kaijser, 2001). To remedy this, and to be able to promote renewable energy, the Ministry for Industry developed new environmental instruments. In 2003 a system of tradable renewable electricity certificates was implemented to encourage the development of renewable energy sources. In this scheme, renewable energy producers were allowed to issue and sell certificates. The retail suppliers and users were obliged to buy a certain amount of these certificates, which would be available on an exchange market. Prices would depend on supply and demand, and so a new type of environmental market was created (Prop, 2002/03:40). In 2005, as the EU moved ahead to fulfil its Kyoto commitments by creating the Emissions Trading System, Sweden was in the front line in developing the Swedish application of this scheme (Prop, 2003/04:132).

In total, these policy developments indicate a shift from the approach taken in 1997. The national-planning and state-building ideology was not applicable in the context of a deregulated and international market. But the prominence of the climate issue grew and grew, and a new type of frame bridging could therefore develop, between energy-as-market and energy-as-risk. In this, the EPI and learning process worked both ways. Swedish energy policy remained faithful to its ultimate objectives in one sense, but shifted gear towards a more internationalized and market-based approach for reaching them. Furthermore, a primacy of the climate issue in energy policy was now achieved, unparalleled in previous rounds.

Framing of Agricultural Policy

Like in most European countries, the history of Swedish agricultural policy follows the transition from a rural to an industrialized society. Its development has its roots in the depression years of 1920–1930, when excess supply and low prices led policymakers to create protectionist policies (Schwaag-Serger, 2001, p83). Intervention measures were introduced and corporatist arrangements set up (Micheletti, 1990; Rothstein, 1992, pp235–239). Although these policies were intended to be temporary, they remained a premise for further developments. The agricultural decision of 1947 set three goals for Swedish agricultural policy: first, a production goal in order to safeguard a secure national supply of food; second, a distribution goal aimed at income parity and stability; and third, an efficiency goal that aimed at production efficiency in order to make high farmer incomes compatible with reasonable consumer prices and overall economic growth. These goals were to be achieved through market regulations and measures aimed at improved production efficiency (SOU, 1946:42). The goals were similar to those later expressed in the European Common Agricultural Policy (CAP). However, the national food supply goal became much more prominent in Sweden than elsewhere in Europe, as self-sufficiency in food was regarded as a necessary part of the Swedish policy of neutrality (Schwaag-Serger, 2001, p84).

The importance of production efficiency was further stressed in the 1960s, partly because the growing industrial sector needed to recruit labour from agriculture. At the same time, urbanization resulted in the problem of keeping people living in sparsely populated rural areas at the borders. The desire to keep a rural population in all parts of the country was rooted in security considerations (again the neutrality policy), even though it developed into an independent goal of regional politics. The 1960 Agricultural Policy Commission that led to the 1967 farm bill thus gave priority to improved production efficiency and reduced the food security goal to 80 per cent. An additional reform was that a consumer delegation was established in farm price negotiations (Flygare and Isacson, 2003, pp233–235). In the 1970s high food prices led to the introduction of retail subsidies. There was shortage of milk, leading to revived concern with food security and rural socio-economic decay. As a result, the 1977 agricultural legislation, in reasserting the priority of income parity and food security, added an explicit consumer goal (Flygare and Isacson, 2003). In the 1980s the debate intensified, and in 1985 the goal of income parity was abandoned, while food supply security was given high priority. However, as the costs for retail subsidies increased dramatically and over-production occurred in certain agricultural sectors, a parliamentary working group for agricultural reform (LAG) was assigned in 1988 to evaluate the 1985 agricultural policy decision, suggest a new policy and analyse state responsibility for surplus areas (Prop, 1988/89:47). The group paved the way for the 1990 agricultural deregulation decision.

Four agricultural policy frames

Throughout this history, four different agricultural policy frames can be discerned. 'Agriculture-for-security' dominated the Swedish policy discourse from the 1947 agricultural decision up until the early 1980s. This frame aims at securing food supply at reasonable prices for the consumer while at the same time ensuring that farmers have incomes comparable to those of industrial workers. In the 1960s, the importance of producing food declined, while there was a growing need for industrial labour. The means of achieving a balance between the different goals was through central government regulation, which, in the corporate tradition of Sweden, was pursued through close cooperation between agricultural organizations and the state (Micheletti, 1990). Production efficiency is necessary in order to maintain a balance between consumer prices and farmer incomes, and environmental measures are generally regarded as costly restrictions on production. Indeed, this frame has traditionally not been characterized by environmental concerns, although security could well involve environmental as well as economic concerns. Agriculture-for-security was initially embraced by most political parties and the farmers' organizations. Although still expressed in policy debates, for example by the Christian Democrats, it has weakened considerably since the end of the cold war.

'Agriculture-as-entrepreneurial-practice' evolved in the 1980s as part of a general shift in policy debate in support of free enterprise and deregulation. Economic experts criticized the notion that agriculture was somehow special and argued that it should be treated like any other kind of industry. Whereas in the 1960s policymakers looked to improved production efficiency in agriculture for the recruitment of industrial labour, in this frame agriculture is seen as a form of (small) enterprise, with the same needs as other businesses to flourish. This is not to say that farmers had not regarded themselves as entrepreneurs before, but the focus shifted in the 1990s. One among several examples is the illustration by the Federation of Swedish Farmers (LRF) of the modern farmer as a man in a black suit with a laptop computer leaning against a tractor. One might expect that a counterpart to the entrepreneurial farmer would be a strong consumer, but the consumer perspective has not, until recently, had real clout in agricultural policy. Agriculture-as-entrepreneurial-practice regards increased environmental demands as restrictions upon production. The costs that arise from environmental requirements must be compensated for. However, what agriculture should produce should also be decided by demand. This means that ecosystem services may be a public good that farmers produce for society in exchange for monetary compensation.

'Agriculture-for-sustainability' emerged at the same time. Problems associated with the use of pesticides and mercury-based fungicides were discovered in the 1960s but not incorporated into agricultural policy until the agri-environmental debate intensified in the early 1970s, when alternative agriculture became recognized. When food surplus became a new problem in the early 1980s, agriculture appeared as a major cause of environmental problems. The 1985 food policy incorporated an action plan to deal with three environmental goals, in addition to the conventional agricultural goals: environmental protection, nature conservation and sustainable agriculture (Vail et al, 1994, p115). Agriculture-for-sustainability addresses three different aspects of sustainability, which have received varying attention over time. One emphasizes sustainable food production and is associated with organic farming and ideas about local foods, embedded, for example, in Agenda 21. The second is concerned with environmental protection, such as eutrophication or traces of pesticides in water. The third concerns nature conservation, especially keeping an open landscape for biodiversity. Whereas the aesthetic dimension of the open landscape was already discussed in the deregulation debate of the 1980s, biodiversity concerns became stronger during the 1990s. A difference between these three aspects is that, although agriculture is an important cause of problems such as eutrophication, production practices such as grazing are also necessary in order to preserve biodiversity. Agriculture can thus be seen as a producer of ecosystem services, a perspective that has encouraged coalitions between environmental and farmers' organizations. While agriculture-for-sustainability contains a complex number of ideas about how different environmental problems are related at different scales, conservation has often dominated public debate. Agriculture-for-sustainability has been endorsed by environmental organizations and, to some extent, the farmers' organizations as a way to market Swedish agriculture (LRF, 1992). During the 1990s, the LRF, sometimes in association with the agricultural authorities, introduced several programmes in order to improve environmental performance on each farm, such as '*Fånga näringen*' ['Catch the Nutrients'] and '*Gårdskontroll*' [Farm control'] (SJV, 2001). The political parties have shown rather low profiles in agricultural policy, even though issues of sustainability have been on the agenda. In the second half of the 1970s the Centre Party was regarded as the green rural party, but the Green Party and parts of the Social Democrats largely took over this role in the 1990s.

'Agriculture-for-development' emerged in the 1970s, when many people thought it was wrong to reduce Swedish food production while world-market prices increased and food scarcity was perceived as a real threat (Flygare and Isacson, 2003, p238). Agriculture-for-development is strongly linked to issues of justice and rights, viewing food security as a fundamental human right (SOU, 1999:78). It regards agriculture as a key sector in development and a possibility for people around the world to secure their livelihoods. In a European and Swedish context, this frame is concerned with rural development, a dimension that has grown stronger in recent years as the CAP has widened its scope. Agriculture-for-development has a local and global, rather than a national, perspective. In the agricultural policy debate, this frame can therefore seem to live a life of its own, evoked in other contexts, such as international aid and regional development policy debates. Environmental issues are connected at both these levels. At the global level, agri-environmental issues ranging from pesticide use to the spread of GMOs are framed as having to do with justice and the empowerment of local communities. In the domestic rural development debate, a clean and attractive environment is regarded as a potential resource for development. Today, the frame is associated with two sets of actors. One consists of development NGOs, including the international activities of major environmental organizations such as SSNC, which often focus on the importance of strengthening local institutions. The other has a more pronounced market orientation, with deregulation of agricultural markets to improve possibilities for third world farmers to export their produce to consumers in countries like Sweden. It can be found in the Swedish International Development Agency and the Ministry of Foreign Affairs.

Table 5.2 *Summary of four agricultural policy frames*

Frame	Agriculture-for-security	Agriculture-as-entrepreneurial-practice	Agriculture-for-sustainability	Agriculture-for-development
Underlying problems	Inadequate supply of food and populated rural areas	Distorted market conditions	Environmental problems from production as well as non-production	Unequal conditions for farmers worldwide, agriculture as basic industry for development
Policy preferences	Price regulations, central interventions to guide development incl. support to improve efficiency	Market instruments, deregulation or ensuring neutral competition conditions	Regulation and compensating ecosystem services	Deregulating markets, supporting local institutions, empowerment
Core values and goals	Ensuring national self-sufficiency in food and thus independence, income parity, production efficiency	Fair competition and conditions for the farm as small enterprise, the farmer as innovative entrepreneur	Environmentally sustainable food production, a diverse rural landscape	Food security for all, (rural) development, local empowerment
Scale view	National	National and European	National and regional	Global and local

Table 5.2 presents some summary features of the four identified agricultural policy frames. The four frames have influenced the extent of EPI in agricultural policy and the formulation of policy since the mid 1980s, as examined below.

The deregulation of agriculture in 1990

Historically, Swedish regulation of agriculture consisted of two elements: border protection against foreign competition and an internal market regulation with negotiated prices. Government subsidies promoted large-scale and specialized practices, leading to environmental degradation (Ds, 1989:63, p259). From 1989, the state also granted direct subsidies in the form of payments for animal or area units, as a result of negotiations within the General Agreement on Tariffs and Trade (GATT) (SOU, 1994:82, p37). The growing costs for society, as well as increasing consumer prices, led to widespread demands for reform of agricultural policies in the 1980s. At the same time, Swedish politicians were confronted with the proposals within GATT for deregulating food trade from 1990. The prospect at the time was that Sweden would remain outside of the EU, but would still be affected by the GATT decisions (SOU, 1994:82, p37). A Swedish food policy reform would adapt Swedish agriculture to the new conditions, leading to a more efficient, more competitive, stream-lined sector. Though initially sceptical, even the LRF was prepared to adopt the liberalization agenda, provided that deregulation took place at a measured pace. In 1989 it gave its public support to proposals which had leaked from the LAG working group (Flygare and Isacson, 2003).

However, environmental concerns were raised about the potential negative impacts of the reform on the landscape, since it was expected to substantially reduce the cultivated agricultural area and have certain regional effects, in particular meaning that remote forested areas would be abandoned and southern agricultural areas would become even more intensively managed. In response, the government substantially increased, as of 1993/1994, the grants directed towards the protection of cultural heritage and landscape values in agriculture. In addition, they launched a programme to reduce nutrient leaching. Farmers who resisted the effort to attune agriculture to the market joined with the wider public who cherished the open landscapes associated with the countryside and with conservation interests. Signs were put up along the roads, asking motorists whether they would prefer to pass dense, dark spruce forests or open meadows (Flygare and Isacsson, 2003). The decision in 1990 to deregulate meant that food prices were determined by the market and not negotiated, and the old goal of income parity was abolished. There was increased attention to qualitative goals such as health and environment (Prop, 1989/90:146). While production subsidies ceased in the deregulation reform in 1990 (Daugbjerg, 1998), new environmental subsidies were introduced in Swedish agriculture in 1986, and these were increased dramatically thereafter (Eckerberg, 1994, p83).

The dominant frame during this round was agriculture-as-entrepreneurial-practice. Farmers who initially endorsed the traditional agriculture-for-security perspective could seek support from agriculture-for-sustainability. Policymakers also learned about the environmental concerns raised, and established environmental programmes to counteract potential negative effects of the reform. However, the focus was still on the external effects of agriculture rather than on environmentally sustainable production.

Entry into the CAP and the Environmental and Rural Development Plan

The transition period came to an end when Sweden joined the EU in January 1995. Despite considerable disagreement among farmers, the LRF supported membership, which would entail reintroducing regulations. Already by the end of 1991, a government committee had been assigned to investigate how Sweden could harmonize its newly adopted food policy to the EU requirements. The result was the introduction of area subsidies, combined with a reduction of Swedish prices for agricultural products (SOU, 1993:33). The EU offered Sweden a special deal, along with the other two new member states Finland and Austria, namely more extensive support to national Environmental and Rural Development Plans (ERDPs) than in the other member states. In the Swedish case, this amounted to 165 million ECU until 1997, to be matched by a national contribution of the same magnitude (SOU 1994:82, p51). Allowing extra subsidies for environmentally friendly agriculture from the EU was a way to compensate for the high membership fee that Sweden faced. At the same time, the government remained concerned with losing perceived benefits from the 'reversed' deregulation.

The first ERDP (1995–1999) focused on environmental issues rather than on rural development. A transition had to be made from earlier national programmes. The main environmental issues to cover were eutrophication, the loss of micro-habitats and organic farming. The regional-political implications of the programme appeared in special support to sugar production on Gotland and the cultivation of brown beans on Öland (these measures have been difficult to abolish despite limited environmental benefits). There was considerable debate on the rather detailed measures designed in the first programme. However, the whole agricultural administration had to find its way in a new context, with new requirements for following up measures (Hagberg, 1996 and 1997).

In connection with Agenda 2000, the European-level CAP reform, the European Council adopted Regulation EC 1257/1999 on support for rural development, concerned with sustainable development of rural areas. This implied that rural development would form a second pillar of the CAP and formed the basis for the second ERDP (2000–2006). It built on an integrated course of action towards rural development in order to preserve the landscape and environment, emphasizing the need to maintain agriculture in areas with particularly difficult conditions. The ERDP adopted two parallel priorities: (1) 'Environmentally Sustainable Agriculture' to conserve resources for future generations and contribute to achieving Swedish environmental quality objectives as well as specific environmental objectives for agriculture; and (2) 'Economically and Socially Sustainable Development' in rural areas requiring a competitive rural economy with production geared towards quality rather than quantity. The measures here would contribute to modernizing and increasing efficiency in agriculture (Eckerberg and Wide, 2001).

The framing of the debate during this period was very pragmatic and to a large extent dominated by the need to adapt quickly to a new administrative system. The process entailed considerable EPI through combining agriculture-for-sustainability with agriculture-as-entrepreneurial-practice and agriculture-for-development. However, the separation of the first ERDP from the ordinary CAP measures structurally cemented the division between the agri-environmental and conventional production policy tracks. At the same time, the Environmental Bill in 1998 introduced sector responsibility for agriculture, along with the environmental objectives (NEQOs) to be implemented in agriculture, and this marked the beginning of a new period in Swedish agricultural policy.

Around the CAP reform in 2003

Triggered by the enlargement of the EU, and the adoption of Agenda 2000 in 1999, the reform of the CAP in 2003 (EC regulation 1782/2003) meant a substantial expansion of agri-environmental measures, requiring member states to make Rural Development Plans with financial support from part of the European Structural Funds and the LEADER programme. Agriculture was seen to promote rural areas (Baldock et al, 2001, p28). Sweden was an active promoter of this change. A parliamentary committee on food and the environment made proposals for a Swedish strategy in the future design of the CAP (SOU, 1997:102). Furthermore, the Minister of Agriculture assigned an interdepartmental working group to investigate how Sweden should implement the new CAP (Skr, 2003/04:137).

The division into two different strands of policy discussions within the EU, the CAP on the one hand and the ERDP on the other, remained an obstacle to a debate on agricultural sustainability covering all three dimensions (social, economic and ecological) and blocked new perspectives on the role of agriculture (Interview, SSNC 2, 2005). The reform implied that all direct support would be decoupled from production. New objectives would include improved competitiveness of farming and forestry, environment and countryside, improving quality of life, and diversification of the rural economy. Instead of the Structural Funds and CAP, only one funding and programming instrument (the European Agriculture Rural Development Fund) would be used, of which a minimum of 25 per cent must be spent in each member state on measures towards the environment and countryside. A national envelope of 10 per cent of the support could be distributed to, for example, environmentally friendly farming (Lundqvist, 2003). In addition, the LEADER fund would support bottom–up initiatives in rural areas.

The Swedish debate concentrated on whether to use a farm model or a regional model for distribution of support and whether and how to use the national envelope. The discussion focused on three goals: fair competition for farmers, maintenance of the open, grazed rural landscape, and fulfilling the NEQOs. The government's decision largely followed the group's proposals except that the coupling to farms was stronger (Ds, 2004:9). Rather than a pure farm or regional model, a mixed model was selected. The right to use a national envelope would be kept, but used restrictively. The funds raised through modulation should be used to strengthen support to less favoured areas, to grazing land and meadows, and to introduce support for growing hay in non-support areas. There would also be support for investment in animal welfare, small-scale food production and a system for agricultural advice (Jordbruksdepartmentet, 2004).

In the debate on how to implement agricultural reform, two frames dominated, agriculture-as-entrepreneurial-practice and agriculture-for-sustainability. Farmer representatives prioritized the stable production conditions that, in reality, also favoured already-active farms. Even though several agri-environmental issues were on the table, the government, along with the environmental NGOs, focused on the protection of biodiversity associated with open landscapes. Despite the fact that the reform entailed increased support for rural development, agriculture-for-development was not very present in the debate. One reason for this might be a highly technical agenda, where choices between different routes of implementation had clear implications for established interests.

Framing of Bioenergy Policy

So far this chapter has studied EPI in two well-defined sectors. Bioenergy policy represents a mix of the energy and agricultural sectors, albeit with a story of its own (see also Chapter 1).

Shifting bioenergy policy frames – Three policy rounds

We can distinguish four bioenergy policy frames: 'bioenergy-for-security', 'bioenergy-as-entrepreneurial-practice', 'bioenergy-for-agricultural-adaptation' and 'bioenergy-for-sustainability'. These are closely related but not identical to the energy and agricultural policy frames explained above. Table 5.3 presents a summary of features of the frames. Three important shifts in bioenergy policy framing have occurred over the investigated time.

Table 5.3 *Summary of four bioenergy policy frames*

Frame	Bioenergy-for-security	Bioenergy-as-entrepreneurial-practice	Bioenergy-for-agricultural-adaptation	Bioenergy-for-sustainability
Underlying problems	Oil-dependency, international competition	Balance security of supply with reasonable prices	Costly surplus agricultural production	Balancing nuclear phase-out with climate policy
Policy preferences	Public investment in domestic energy production and agricultural marginal protection	Market instruments and competition, minimal public interventions	Public investment in agricultural conversion to alternative production	Public investment in research, production and use of renewable energy
Core values and goals	Conservation, welfare state-building, employment, welfare creation, national efforts, secure energy/food supply	Market-economy, rural employment, consumer focus, international efforts, cheap energy and food	Rural landscape conservation, national efforts, agricultural adaptation to market conditions	Ecologism, resource efficiency, international efforts, sustainable energy supply and agriculture
Scale view	National (European) and long term	International and long term	National (European) and short term	International and long term

Source: adapted from Söderberg (2005)

The emergence of bioenergy rhetoric

Bioenergy-for-security dominated the 1970s and early 1980s. At that time, Sweden was moulded by the cold war. The priority was to provide a credible basis for the Swedish neutrality policy by securing a domestic supply of food and energy. The oil crisis of 1973–1974 had triggered a political awakening concerning the vulnerability of the Swedish energy supply (70 per cent of the total Swedish energy supply came from oil), which brought about the overarching goal for energy policy during this period of time: a secure energy supply and nuclear phase-out through increasing the use of domestic, preferably renewable, energy sources (SOU, 1978:17; Prop, 1979/80:170 and 1980/81:49). Bioenergy, especially energy crops, first became interesting for policymakers in the 1970s as it offered a solution to two fundamental policy problems. First, bioenergy could reduce Sweden's oil-dependency and vulnerability to fluctuations in international oil prices. Second, cultivation of energy crops could help curb the costly agricultural surplus (Prop, 1985/86:74; DsJo, 1986:6 and 1987:3). After the 1979 Three Mile Island incident and the second oil crisis in 1981, an ambitious programme was introduced to replace oil, including investments in solar energy and biomass. Consequently, there was a rapid expansion of district heating plants. Together with high oil prices and simultaneous calls for alternative production of biomass on surplus agricultural land, this gave bioenergy a good market position during 1980–1985 (see Chapter 1). Apart from some concerns raised over the possible visual effect on the rural landscape of the large-scale cultivation of energy forest[2] on agricultural land, the environment was not an issue. Bioenergy-for-security was essentially unchallenged until the 1980s.

Nuclear phase-out and low-priced oil shifted focus away from long-term development of new energy sources to rapid introduction of domestic fuels in the 1980s. It became increasingly important to balance security of supply with reasonable prices and more focus was placed on the consumer. In short, a new frame, bioenergy-as-entrepreneurial-practice, emerged. Closely related to energy-as-market, it emphasized that bioenergy production constitutes a cheap and efficient energy supply but also presented an opportunity for rural job creation and entrepreneurship. In considering environmental arguments, competition between different renewable energy sources to accomplish the highest possible efficiency was considered (at least) equally important. The main supporters of bioenergy-as-entrepreneurial-practice were similar to the supporters of energy-as-market: the Conservative and Liberal parties, academic economists, energy companies, the Ministries of Finance and Industry and the Swedish Energy Agency. The Green Party today partly embraces this frame (in a 'green variant'), arguing for environmentally friendly bioenergy production as a potential Swedish growth sector.

In 1987 the government dramatically changed the preconditions for agricultural policy, opening the Swedish food supply system to increased imports. The government had noticed the negative effects of food price regulation and was influenced by discussions on liberalization in the OECD and GATT talks (SOU, 1987:44). The proponents of the emerging bioenergy-for-agricultural-adaptation advocated a controlled deregulation of agriculture, upholding marginal protection and providing state support for shifts from food production to energy crop production. Basically, this frame also signalled the LRF's reaction to the changing conditions for agriculture under the market economy. As long as agriculture preserved an open landscape, any extra environmental measures imposed on agricultural practices were seen as unnecessary. The LRF still partly lingers in bioenergy-for-agricultural-adaptation, advocating economic support to farmers to increase bioenergy production, although today strongly influenced by environmental and entrepreneurial arguments.

In short, from initially being a question of national security and of interest to both the energy and the agricultural sectors, the energy sector's view on bioenergy-as-entrepreneurial-practice, combined with the agricultural bioenergy-for-agricultural-adaptation, transformed energy crop production into mainly an agricultural issue during the mid to late 1980s.

Agricultural deregulation and the 1991 energy agreement

Bioenergy-for-sustainability started to surface in the 1990s. Bioenergy now became interesting because of its environmental benefits in the light of the nuclear phase-out and climate concerns. Environmental NGOs, the Green and Centre parties, parts of the Social Democrats as well as the Swedish Environmental Protection Agency (SEPA), the Ministry for the Environment (MoE) and gradually also the LRF saw an opportunity to argue for support to renewable energy sources, including bioenergy production in agriculture. Two groundbreaking bills were presented in 1990 and 1991, signifying a breakthrough for environmental policy integration in Swedish energy policy and agricultural policy. The new policy led to an increase in both bioenergy research and production volumes during the years to come. First, the new food policy (Prop, 1989/90:146) implied deregulation of the agricultural sector and a reframing of agriculture to make it important not only for food production, but also for domestic production of environmentally friendly energy. Energy crops were an important part of the solution to help the agricultural sector adjust to the new conditions. The main argument for their cultivation was to stimulate alternative uses of farmland (bioenergy-for-agricultural-adaptation) (Ds, 1989:63, p215). Although environmental arguments were present, market issues dominated the reasoning for increased energy crop production, which incorporated bioenergy-as-entrepreneurial-practice. Second, the 1991 energy agreement (Prop, 1990/91:88) paved the way towards a new, more environmentally sound energy system. Policies aimed for the successive commercialization of bioenergy through research, investment programmes and tax-exemption-bridged agendas from bioenergy-as-entrepreneurial-practice and bioenergy-for-sustainability.

As a result of environmental awakening in the years that followed, sustainability issues took their place in bioenergy rhetoric as well as in strategies, more markedly in the energy sector but also visibly in the agricultural sector (SOU, 1992:90). The framing of bioenergy shifted to combining bioenergy-for-agricultural-adaptation and bioenergy-as-entrepreneurial-practice with bioenergy-for-sustainability. There was an agricultural policy objective to adjust to new market conditions and an energy policy objective to phase out nuclear power and increase the use of renewable energy. Furthermore, the environmental goals of energy policy were incorporated into agricultural policy. These factors led to coordinated policy goals and strategies aiming for the same targets. When EPI in the two sectors was combined with horizontal coordination, one practical outcome was a rapid expansion of forest cultivation for energy.

Generally speaking, this frame change was spurred partly by the international attention to environmental issues springing from the Brundtland Report (see Chapter 1) and partly by the political conflict that followed in the wake of the 1988 decisions on locking CO_2 emissions to the present level and kick-starting the nuclear phase-out (Prop, 1987/88:90). The need for a revision of Swedish energy policy became apparent around the same time that the quest for agricultural deregulation was growing stronger due to rising costs for price and export support. The boost for agricultural bioenergy production in the early 1990s can also partly be explained by the LRF's influence on agricultural policy, since they took a very critical position on the process preceding deregulation.[3] This may have improved their influence in the process: the chairman of the LRF was able to close a deal that compensated the agricultural sector for the changed policy and bought the sector time for a transition to a more market-based situation including production of energy crops (Flygare and Isacson, 2003; Söderberg, 2005).

CAP implementation and the 1997 energy agreement

Bioenergy-for-sustainability further strengthened during the 1990s. The energy transition to non-fossil and non-nuclear energy was seen as the key obstacle in creating the 'green people's home' (see Chapter 1). Simultaneously, however, the rhetorical attention to a restructured agricultural production decreased. Forest crop cultivation for energy on agricultural land, which had previously had a prominent position in Swedish bioenergy policy (nearly the entire planting cost was covered by subsidies, which brought about a steady rise in energy forest cultivation during 1990–1996 (STEM, 2003)), suddenly found itself in the backwaters of agricultural and energy policy. Instead, other renewable fuels or semi-renewables, such as waste incineration, gained ground in the process leading up to the energy policy bill 'A sustainable energy supply' (Prop, 1996/97:84).

Bioenergy in general was now propounded on the basis of sustainability, and possible environmental consequences from bioenergy utilization were also taken into consideration in Swedish energy policy. Bioenergy production on farmland was promoted on the basis of entrepreneurial (rural job creation) as well as sustainability arguments (Prop, 1997/98:2 and 1997/98:145; Skr, 1998/99:5). Agricultural policy recognized energy policy goals and a reframing towards sustainability was taking place in bioenergy policy, especially due to climate change concerns. Still, the maximum economic support allowed within the CAP constrained the possibilities for the government to stimulate increased cultivation of energy forest on Swedish farmland. Drastic subsidy reductions in 1997 led to an immediate and abrupt drop in energy forest planting (STEM, 2003). Markedly, and despite the aim of formulating an agricultural policy promoting the conversion of Sweden into an ecologically sustainable society, neither the Environmental and Rural Development Plan for Sweden (Jordbruksdepartementet, 2000) nor its preceding committee (SOU, 1999:78) mentioned agricultural biomass production. Bioenergy received rhetorical attention based on bioenergy-as-entrepreneurial-practice and bioenergy-for-sustainability in both sectors, but in practice it became an energy sector issue after the Swedish implementation of the CAP. The coordination deficiency between Swedish energy policy goals and the CAP limited the possibilities for national policy coordination regarding energy forest cultivation.

Bioenergy policy in the new millennium

The new millennium brought with it a new framing of Swedish bioenergy policy. From the systematic move towards sustainability during the 1990s, the sustainability frame lost its dominance. Paradoxically, however, the ambitions of renewable energy were concurrently raised. In the 2002 energy policy decision the supply was set to increase to 10TWh in 2010, although not only, or even primarily, on environmental grounds: the climate issue was only mentioned as the third of three reasons (Prop, 2001/02:143). Instead, supply security (driven also by the EU) and nuclear phase-out appeared as the main reasons for the renewable energy venture. In the mid 2000s, supply security has risen to the top of the agenda not only in Sweden but throughout the EU and globally. This has been triggered by high oil prices, increasing political unrest in the oil-producing countries and sudden events like the break in gas supply from Russia due to a disagreement between Russia and the Ukrainian. The policy course undertaken within Sweden has simultaneously shifted to a more market-oriented approach: from supply stimulation of energy crop cultivation to demand stimulation through the introduction of electricity certificates.

Electricity production from wind, solar, geothermal, some hydro and some biofuels was initially defined as energy that should be valid for certificates (Prop, 2002/03:40). In 2004 the government included peat fuels, which many analysts consider a non-renewable resource (Prop, 2003/04:42). Thus bioenergy policy has mainly turned into an energy-sector issue and taken a turn towards emphasizing market and cost-efficiency (bioenergy-as-entrepreneurial-practice) over environment (bioenergy-for-sustainability) during the last few years.

Conclusions

Both energy and agricultural sectors have undergone serious reframing processes during the time period examined. In the energy sector, energy-as-infrastructure has given way to energy-as-market and, through the influence of energy-as-risk, EPI has occurred. In particular, the climate issue has become institutionalized and an integral part of the dominant frame. This must be considered a significant EPI process over the last 15 years, although it appears to have crowded out other important environmental issues associated with energy. Agricultural policy since the 1990s has evolved from deregulation of agriculture in 1990, through re-introduced regulation in 1995 with the CAP and the first environmental and rural development programme, to the CAP reform beginning in 2003. Over this time, agriculture-for-sustainability has gradually gained ground, in particular at the expense of agriculture-for-security. In the agricultural sector, a broader set of environmental concerns has become central in policies. Heavy significance is put on landscape issues, in terms of cultural and natural values, as well as management of nutrient leakage. In addition, support to ecological production has increased. However, the environmental issues have become integrated by way of a particular policy track, the rural development programme, whereas the policy mainstream, concerned with agricultural production, remains relatively unconcerned with environmental issues. The lack of coordination between the two sectors has also constrained the potential for bioenergy, which remains at the crossroads of the two sectors and depends on frame alignment across sectors, rather than within sectors. Recent years have manifested less policy coordination between the agricultural and energy sectors regarding bioenergy production, even though EU policy has changed to include increased support for energy crop cultivation in agriculture. It also seems that the EU has brought back bioenergy-for-security to energy policy, agricultural policy and bioenergy policy.

The evolution of the framing in the three sectors displays some commonalities. First, they follow the internal logic of the development of the respective sector. By this we mean that the policy framing, to a large extent, is an adaptation to driving forces such as international policy streams, actor changes and market developments. Second, the 'set' of competing frames display similarities in several principal dimensions that clearly relate to larger dynamics of post-war European politics. Premises such as national security, public sector expansion, market orientation and post-normal critiques show up across the sectors. Third, trend breaks in framing occur within similar intervals and often at similar points in time. There are clearly broader logics of politics that have a large influence on the occurrence of windows for policy learning and reframing. At the same time, there are also relatively important differences between the cases, such as which environmental issues are given attention, which learning processes sectoral actors have engaged in and in what ways the environmental dimensions manifest themselves in the sectoral policy. To unpack these questions further, Chapter 6 looks more deeply into the institutional characteristics of the sectoral policy machinery and to what extent they can help us explain the differential patterns of EPI.

Notes

1   During the cold war, Sweden pursued a neutrality policy, aiming to be neutral in the case of a war between the superpowers. When the cold war ended, and the Soviet Union dismantled, the neutrality policy had played out its role. Sweden has, however, continued a policy of military non-alliance.

2   Energy forest, also referred to as *Salix* or short rotation coppice (SRC), is an energy crop consisting of fast growing trees that can be cultivated on agricultural land and used in energy (electricity, heat and fuel) production. Other kinds of energy crops are energy grass, cereals and oil-yielding plants.

3   A broad range of organizations representing environmental, agricultural, regional and defence interests were consulted in the referral process on the LAG committee proposal in the run-up to the new food policy bill. Of these, the LRF took a very critical standpoint, which may have improved their negotiation position in the process between the committee proposal and the actual bill. Apart from criticizing the lack of expert groups and the consequently insufficient analysis, the LRF claimed that their representatives had not been allowed to participate to a full extent in the working group meetings. Therefore, the LRF's viewpoint was that a decision on future Swedish food policy could not rest on the LAG proposal, but rather should be based on the LRF guidelines (Prop, 1989/90:146, attachment 2).

References

Anshelm, J. (2000) *Mellan frälsning och domedag: Om kärnkraftens politiska idéhistoria i Sverige 1945–1999 [Between Salvation and Armageddon: On Nuclear Political History in Sweden 1945–1999]*, Brutus Östlings Bokförlag, Stockholm

Baldock, D., Dwyer, J., Lowe, P., Petersen, J.-E. and Ward, N. (2001) *The Nature of Rural Development: Towards a Sustainable Integrated Rural Policy in Europe*, WWF and the Great Britain Countryside Agencies, IIEP, London

Bergman, L. (2001) 'The economics of Swedish energy policy', in S. Silveira (ed) *Building Sustainable Energy Systems: Swedish Experiences*, AB Svensk Byggtjänst and Swedish National Energy Administration, Stockholm

Daugbjerg, C. (1998) *Policy Networks under Pressure: Pollution Control, Policy Reform and the Power of Farmers*, Ashgate Publishing, Aldershot, UK

Ds (1989:63) 'En ny livsmedelspolitik' ['New food policy'], Ministry Publication Series, Regeringskansliet, Stockholm

Ds (2001:60) 'Effektivare energianvändning' ['More efficient energy use'], Regeringskansliet, Stockholm

Ds (2004:9) 'Genomförandet av EU:s jordsbruksreform i Sverige ['The implementation of CAP reform in Sweden'], Regeringskansliet, Stockholm

DsJo (1986:6) 'Åtgärder för att minska spannmålsöverskottet. Rapport 2 från spannmålsgruppen' ['Measures to reduce cereals surplus. 2nd Report from the cereal group'], Ministry Publication Series, Regeringskansliet, Stockholm

DsJo (1987:3) 'Intensiteten i jordbruksproduktionen. Miljöpåverkan och spannmålsöverskott' ['The agricultural production intensity. Environmental impact and cereals surplus'], Ministry Publication Series, Regeringskansliet, Stockholm

Eckerberg, K. (1994) 'Consensus, conflict or compromise: The Swedish case', in K. Eckerberg, P. Mydske, A. Iilahti and K. Pedersen (eds) *Comparing Nordic and Baltic Countries: Environmental Problems and Policies in Agriculture and Forestry*, Nordic Council of Ministers, Copenhagen

Eckerberg, K. and Wide, J (2001) 'The nature of rural development', Report 2001:1, Department of Political Science, Umeå University, Umeå, Sweden

Engström, R., Nilsson, M. and Finnveden, G. (2006) 'Issue characteristics and policy attention in two Swedish sectors: Agriculture and energy', submitted manuscript

Flygare, I. and Isacson, M. (2003) *Jordbruket i välfärdssamhället [Agriculture in the Welfare Society]*, Natur och Kultur, Stockholm

Hagberg, L. (1996) 'Politics of harmonisation? Swedish EU membership and changing use of agri-environmental policy instruments', European Policy Process Occasional Papers, Department of Political Science, University of Essex, UK

Hagberg, L (1997) 'The reception of EU agri-environmental policy at the local level – The case of the Laholm Bay area', European Policy Process Occasional Papers, Department of Political Science, University of Essex, UK

Hajer, M. (1995) *The Politics of Environmental Discourse: Ecological Modernization and the Policy Process*, Oxford University Press, Oxford, UK

Holmberg, S. and Weibull, L. (2006) *Du stora nya värld: SOM-undersökningen 2005 [You Great New World: The SOM Poll 2005]*, Gothenburg University, Gothenburg, Sweden

Jordbruksdepartementet (2000) 'Förordning 2000:577 Miljö och landsbygdsutvecklingsprogrammet 2000–2006' [Environmental and rural development plan for Sweden 2000–2006'], Regeringskansliet, Stockholm

Jordbruksdepartementet (2004) 'Genomförandet av EU:s jordbruksreform i Sverige' ['The implementation of the EU agricultural reform in Sweden'], factsheet, Regeringskansliet, Stockholm

Kaijser, A. (2001) 'From tile stoves to nuclear plants – The history of Swedish energy systems', in S. Silveira (ed) *Building Sustainable Energy Systems: Swedish Experiences*, AB Svensk Byggtjänst and Swedish National Energy Administration, Stockholm

LRF (1992) *Sveriges bönder – steget före: Konsumentpolitiskt handlingsprogram [Swedish Farmers – One Step Ahead: Action Programme for Consumer Policy]*, LRF, Stockholm

Lundgren, L. (1978) *Energipolitik i Sverige 1890–1975 [Energy Policy in Sweden 1890–1975]*, Sekretariatet för framtidsstudier, Stockholm

Lundqvist, L-E. (2003) 'Innehållet i reformen' ['Content of the reform'], *Internationella perspektiv. Nyhetsbrev från LRF*, vol 21, pp4–5

Micheletti, M. (1990) *The Swedish Farmers' Movement and Government Agricultural Policy*, Praeger, New York

Nilsson, M. (2005) 'Learning, frames and environmental policy integration: The case of Swedish energy policy', *Environment and Planning C: Government and Policy*, vol 23, pp207–226

Prop (1979/80:170) 'Om vissa energifrågor' ['On certain energy issues'], Government Bill, Regeringskansliet, Stockholm

Prop (1980/81:49) 'Stöd för åtgärder att ersätta olja m.m.' ['Support for oil-replacing measures'], Government Bill, Regeringskansliet, Stockholm

Prop (1980/81:90) 'Om riktlinjer för energipolitiken' ['On directions for energy policy'], Government Bill, Regeringskansliet, Stockholm

Prop (1985/86:74) 'Jordbruksforskning m.m.' ['Agricultural research and more'], Government Bill, Regeringskansliet, Stockholm

Prop (1987/88:90) 'Om energipolitik inför 1990-talet' ['On energy policy for the 1990s'], Government Bill, Regeringskansliet, Stockholm

Prop (1988/89:47) 'Om vissa ekonomisk-politiska åtgärder, m.m.' ['On certain economic-political measures etc'], Government Bill, Regeringskansliet, Stockholm

Prop (1989/90:146) 'Om livsmedelspolitiken' ['On food policy'], Government Bill, Regeringskansliet, Stockholm

Prop (1990/91:88) 'Om energipolitiken' ['On energy policy'], Government Bill, Regeringskansliet, Stockholm

Prop (1996/97:84) 'En uthållig energiförsörjning' ['A sustainable energy supply'], Government Bill, Regeringskansliet, Stockholm

Prop (1997/98:2) 'Hållbart fiske och jordbruk' ['Sustainable fisheries and agriculture'], Government Bill, Regeringskansliet, Stockholm

Prop (1997/98:145) 'Svenska miljömål. Miljöpolitik för ett hållbart Sverige' ['Swedish environmental objectives: Environmental policy for a sustainable Sweden'], Government Bill, Regeringskansliet, Stockholm

Prop (2001/02:143) 'Samverkan för en trygg, effektiv och miljövänlig energiförsörjning' ['Cooperation for a safe, efficient and environmentally friendly energy supply'], Government Bill, Regeringskansliet, Stockholm

Prop (2002/03:40) 'Elcertifikat för att främja förnybara energikällor' ['Electricity certificates to promote renewable energy sources'], Government Bill, Regeringskansliet, Stockholm

Prop (2003/04:42) 'Torv och elcertifikat' ['Peat and electricity certificates'], Government Bill, Regeringskansliet, Stockholm

Prop (2003/04:132) 'Handel med utsläppsrätter' ['Trade with emissions rights'], Government Bill, Regeringskansliet, Stockholm

Rothstein, B. (1992) *Den korporatistiska staten: Intresseorganisationer och statsförvaltning i svensk politik [The Corporatist State: Interest Organizations and State Authority in Swedish Politics]*, Norstedts juridik, Stockholm

Schwaag-Serger, S. (2001) *Negotiating CAP Reform in the European Union: Agenda 2000*, Swedish Institute for Food and Agricultural Economics, Lund, Sweden

SJV (2001) 'Projektplan för Greppa Näringen' ['"Catch the Nutrients" project plan'], 26 February

SJV (2006) 'Bioenergi – Ny energi för jordbruket' ['Bioenergy – New energy for agriculture'], Rapport 2006:1, Jordbruksverket, Jönköping, Sweden

Skr (1998/99:5) 'Hållbara Sverige – Uppföljning och fortsatta åtgärder för en ekologiskt hållbar utveckling' ['Sustainable Sweden – Follow-up and continued measures for an ecologically sustainable development'], Government Communication, Regeringskansliet, Stockholm

Skr (2003/04:137) 'Genomförande av EU:s jordbrukspolitik i Sverige' ['Implementation of EU agricultural policy in Sweden'], Government Communication, Regeringskansliet, Stockholm

Söderberg, C. (2005) 'Much ado about nothing? Energy forest cultivation in Sweden: On policy coordination and EPI in a multisectoral issue', paper presented at the ISA RC 24 conference 'Double Standards and Simulation: Symbolism, Rhetoric and Irony in Eco-Politics', University of Bath, 1–4 September

SOU (1946:42) 'Riktlinjer för den framtida jordbrukspolitiken' ['Guidelines for future food and agricultural policy'], Government Committee Report, Regeringskansliet, Stockholm

SOU (1978:17) 'Energi. Betänkande av energikommissionen' ['Energy. Report from the Energy Commission'], Government Committee Report, Regeringskansliet, Stockholm

SOU (1987:44) 'Livsmedelspriser och livsmedelskvalitet. Betänkande av 1986 års livsmedelsutredning' ['Food prices and food quality. Report from the 1986 food investigation'], Government Committee Report, Regeringskansliet, Stockholm

SOU (1990:21) 'Den elintensiva industrin under kärnkraftsavvecklingen' ['Electricity-intensive industry under nuclear phase-out'], Government Committee Report, Regeringskansliet, Stockholm

SOU (1992:90) 'Biobränslen för framtiden' ['Biofuels for the future'], Government Committee Report, Regeringskansliet, Stockholm

SOU (1993:33) 'Åtgärder för att förbereda Sveriges jordbruk och livsmedelsindustri för EG: Betänkande av Omställningskommissionen' ['Measures to prepare Sweden's agriculture and food industry for the EU entry: Report of the commission for EU adaptation'], Government Committee Report, Regeringskansliet, Stockholm

SOU (1994:82) 'Förstärkta miljöinsatser i jordbruket: Svensk tillämpning av EG:s miljöprogram' ['Increased environmental instruments in agriculture: Swedish implementation of the EC environmental programme'], Government Committee Report, Regeringskansliet, Stockholm

SOU (1995:139–140) 'Omställning av energisystemet' ['Restructuring the energy system'], Government Committee Report, Regeringskansliet, Stockholm

SOU (1997:102) 'Mat och miljö: Svensk strategi för EG:s jordbrukspolitik i framtiden' ['Food and Environment: A Swedish strategy for future EC agricultural policy'], Government Committee Report, Regeringskansliet, Stockholm

SOU (1999:78) 'Jordbruk och miljönytta' ['Agriculture and environmental benefits'], Government Committee Report, Regeringskansliet, Stockholm

SOU (2003:80) 'EFUD – En del i omställningen av energisystemet' ['EFUD – One part of the restructuring of the energy system'], Government Committee Report, Regeringskansliet, Stockholm

SSNC (2003) *När larmen tystnar [Silence of the Alarms]*, Svenska Naturskyddsföreningen, Stockholm

STEM (2003) *Uppdrag att utvärdera förutsättningarna för fortsatt marknadsintroduktion av energiskogsodling [Mission to Evaluate the Preconditions for Continued Market Introduction of Energy Forest Cultivation]*, Statens Energimyndighet, Eskilstuna, Sweden

Vail, D., Hasund, K.-P., Drake, L. (1994) *The Greening of Agricultural Policy in Industrial Societies: Swedish Reforms in Comparative Perspective*, Food Systems and Agrarian Change, Cornell University Press, Ithaca, NY, US

Vedung, E. (1982) 'Energipolitiska utvärderingar 1973–81' ['Energy policy evaluations 1973–81'], Delegationen för energiforskning [Delegation for Energy Research], Stockholm

Vedung, E. and Brandel, M. (2001) *Vattenkraften, staten och de politiska partierna [Hydropower, the state and the political parties]*, Nya Doxa, Nora, Sweden

6
Institutional Analysis of Energy and Agriculture

Katarina Eckerberg, Måns Nilsson,
Åsa Gerger Swartling and Charlotta Söderberg

While Chapters 4 and 5 studied environmental impacts and policy framing towards EPI, this chapter will explore the institutional arrangements. This involves studies of how the sectors are organized and how their institutional procedures have evolved as new policies have been prepared and policy frames have changed. As in Chapter 5, we examine the agriculture and energy sectors, with bioenergy as a cross-cutting case, and analyse how their institutions have contributed to, or constrained, EPI progress. The major policy rounds already identified from the late 1980s, mid 1990s and early 2000s serve as a basis for comparison to evaluate the influence of institutional factors across sectors.

The framework for performing the institutional analysis is organized around the following themes and questions. First, we will concentrate on how organizational structures and both formal and informal rules influence the potential for EPI. Here we discuss to what extent norms and power relations within the sectors affect the ways that they interact with various interests and actors outside, including the influence of party politics. It is also necessary to evaluate whether the sector has been defined in the same way over time, since it is clear from the analysis in Chapter 5 that policy frames have indeed changed over time. Are such changes related to organizational changes, such as the involvement of particular actors with certain interests? Or are they spurred by new trends in public opinion, accompanied by changing value systems within organizations? Have special co-ordination units or working groups been designed to promote cooperation between different government departments and agencies and with non-government actors including business and interest groups?

Second, we will investigate the extent to which procedural arrangements and the acquisition of knowledge in decision-making processes influence EPI in the respective sectors. What kinds of deliberative, communicative and dialogue processes have taken place? What procedures were used, who was invited into the processes and how were different alternatives assessed? To what extent were environmental aspects of policy discussed? With a rationalistic view of policymaking, we may assume that EPI as a continuous learning process is more likely to occur if a range of different interests are allowed in, and if competing goals and alternative solutions are weighed and discussed. For instance, the national environmental quality objectives (NEQOs) might play a role here in setting priorities and raising awareness about environmental effects. Or there might be other mechanisms in place on a regular or irregular basis, such as environmental management systems, accounting systems or impact assessment. Thus we examine what kinds of planning instruments are used and discuss the extent to which such instruments contribute to EPI.

Energy Policy

Chapter 5 analysed patterns of reframing and EPI in three energy policy rounds. The first round surrounded the government bill 'Energy policy for the 1990s', which introduced 'energy-as-risk' (Prop, 1987/88:90). The second round surrounded the 1997 energy policy agreement where nuclear phase-out became practical policy and a transition to a renewable energy system became mainstream (Prop, 1996/97:84). The third round took place in the early 2000s, featuring reassessments of governmental policy in the face of a deregulated electricity market and increasing competition, and integration of international climate policy (Prop, 2001/02:55, 2003/04:132). In this section, we examine the evolution of institutional aspects over time and across these rounds and discuss in which ways this has affected the EPI potential in the energy sector.

Institutional and organizational landscape

During the last two decades, the institutional landscape of energy policy has shifted dramatically. The overarching aspect is the institutionalization of the market orientation of the sector, entailing deregulation of distribution, changes in owner constellations among energy industries, such as privatization of municipal and other energy companies, and subsequent consolidation and internationalization. But the most important institutional change was the creation of the Nordic electricity market, under preparation for many years and taking effect in the late 1990s. Interestingly, this development was detached from other strands of energy policy development, such as environmental and safety-of-supply considerations, and the implications of the integrated market seemed to take people by surprise. For instance, once the market was up and running, policies such as production taxes on fossil-fuelled heat and power production became ineffective. This was because the cost increase from taxation only triggered instantaneous production shifts abroad. However, as these patterns became evident to policymakers, the government adjusted relatively quickly, and applied new instruments such as renewable electricity certificates (Prop, 2002/03:40). These large-scale changes thus inspired policy learning in a general sense, providing a window to reflect and reassess certain assumptions. As reflections also tended to include sustainability-related issues, we also saw partial EPI. EPI was not causally related to the market shift, but it was at least partly triggered by the major change in the institutional landscape.

Organizations directly controlled by the government have also undergone radical shifts. In 1992 the governmental agency Vattenfallsverket was incorporated into a state-owned limited company. This changed their operations as well as the political influence on their activities. Essentially, Vattenfall AB has been run as a standard multinational for-profit company, with few considerations beyond the 'bottom line' (Riksrevisionen, 2004b). This has arguably led to a less ambitious environmental agenda, as indicated by their recent strategy of becoming a major European energy player by, for instance, expanding into lignite production, which is harmful to the environment both on- and off-site. In 1997 the parties to the 1997 agreement claimed that we needed a new champion to execute the transition of the energy system, and created a new national energy agency (STEM), shifting the mandate over energy issues from the agency for technology and business development (NUTEK).

Within the government offices, the mandate over energy policy has moved back and forth between ministries several times. Until 1988 it belonged in the Ministry for Industry (MoI). Between 1988 and 1990 it resided within the new Ministry for the Environment and Energy. In 1991 it was put back with the MoI. The MoI's mandate expanded in the late 1990s to form a 'super-ministry' for industry, transport and communications. However, in 2005, the energy section had to move again when the Prime Minister announced the formation of a new Ministry for Sustainable Development, basically expanding the scope and mandate of the Ministry for the Environment (MoE). Up until this change, renewable energy was an MoI concern, whereas climate policy belonged to the MoE. With this shift, energy and climate became handled in the same ministry. With the shift in government to a Liberal–Right coalition in 2006, energy issues again moved back to the MoI. However, the EPI implications of these shifts between ministries seem limited; perhaps due to the relatively effective 'joint drafting' procedure in place between ministries, there is a fair amount of coordination going on in any case.

Corporatism is a main feature of Swedish policymaking. A relative weakening of the corporatist structures noted in Chapter 5 appears to have had mixed, but largely positive, effects. Initially the effect was positive, as the internationalization and liberalization of energy systems contributed to a breaking up of the relatively closed, corporatist network that shaped energy policy up to the 1980s (see next section). As a result, new actors flowed in: academics, environmentalists and niche-industry actors. 'Energy-as-risk' expanded and became a part of the sector. Once the environmental interests became established in the governmental system, it appears that EPI in relation to climate change was facilitated. Nevertheless, we can also observe the effect that many pertinent environmental issues, for example landscapes and air pollution, have dropped off the agenda because they have no established interest groups on the inside advocating their role. Hence, whereas climate issues are well integrated in energy policy, few other environmental issues are considered to any significant degree.

A permanent institutional feature is the split between policy preparation through committees of inquiry and negotiations. In line with the general Swedish tradition, both parliamentary committees and expert committees have been established to contribute to energy policy formation and analyse its impacts. These have contributed to policy learning. In parallel, however, political negotiations have been ongoing, since the Social Democrats have relied on support from other parties to pass bills through parliament. For instance, the 1991 energy policy bill, which revoked the former decision to close the two reactors and cancel the climate policy target, was based on a tri-party agreement between Social Democrats, the Centre and the Liberals. A facilitating factor for the agreement was the expected change in government in 1991 – all polls pointed towards a Conservative–Centre–Liberal coalition and the Centre Party, which had invested in an anti-nuclear position in opposition to its coalition partners, needed to get the issue out of the way and off the agenda. Another example is the 1997 agreement reached between the Social Democrats, the Centre and the Socialists. This was a very strong deviation from the conclusions of the preceding 'Energy Commission', which had negotiated an agreement between the Social Democrats, the Liberals and the Christian Democrats (SOU, 1995:139–140). Thus there are parallel tracks of learning, political discourses and negotiations that do not necessarily communicate with each other very much. The importance of the Prime Minister and his dealings with other parties at the highest political level shines through – weakening the role of the formal institutional processes such as committees, which in principle could be key vehicles for policy learning and for EPI.

Actors' access to policymaking

The relative strength and access of different actors in the energy-policy arena has shifted as a result of contextual and institutional changes. The energy sector was traditionally dominated by industry, such as power producers, equipment providers and energy-intensive industrial users, forming an 'iron triangle' with the labour union and the government, and defining the policy agenda. The fact that the nuclear expansion continued in full force several years after it was decided through the referendum to phase out nuclear power can be taken as an indication of the vested power of this network (Kaijser, 2001). However, as a result of internal changes to the sector, where major industries such as ASEA were transformed into international corporations (ABB), in combination with external pressures, whereby environmental and nuclear risk issues were firmly brought to a head, the network weakened (see Chapter 5). This opened the policy process to environmental interests and to niche players dealing with renewable energy, which had a powerful effect on the energy policy agenda. Besides adherents inside the government, some prominent energy experts and scientists, the Head of Administration of the Swedish National Energy Administration (STEM) and environmental NGOs such as Folkkampanjen (the Public Campaign) and SSNC (the Swedish Society for Nature Conservation) lobbied for environmental protection and safety. As a result, in 1988 the Social Democrats could present a relatively radical energy policy bill, requesting a redirection in energy policy from an expansionist paradigm towards nuclear phase-out and energy savings (Prop, 1987/88:90).

However, as we saw in Chapter 5, in a relatively short time period, a more industrial-oriented perspective of national energy policy became dominant. This change in policy coincided with the Right–Liberal coalition gaining political power (the Moderate Party, the Liberals, the Centre and the Christian Democrats over the period 1991–1994). Yet the shift was far from being simply a consequence of the change in government, but more broadly a trend in conjunction with massive resistance from powerful actors and network groups. In fact, the 'new' approach to energy policy in 1988 had met immediate and strong opposition from various powerful stakeholder groups both within and outside the Social Democrats. In particular, there was strong opposition from their traditionally close ally, the Swedish Trade Union Confederation (LO), which has great influence over the Social Democratic Party. Furthermore, representatives from the power industry, academics, and representatives from major industries interfered in the political debate in an attempt to put pressure on the government to revoke the nuclear phase-out programme (Anshelm, 2000). Eventually, the nuclear phase-out decision became so politically controversial that not only was the programme revoked, but so were most environmental constraints set on energy in the 1988 bill. In addition, the Prime Minister decided to move the energy policy mandate back to the MoI, replacing Birgitta Dahl with the former Deputy Chairman of the LO (Rune Molin) as Minister for Energy in January 1991 (Radetzki, 2004). The advocates for nuclear phase-out had been, albeit temporarily, defeated. Although the old network had been weakened, this time it mobilized its resources successfully. Still, the general opening of the sector was a seed of change that later induced massive changes in policy framing and agendas. The opening of the sector was a necessary, but insufficient, condition for EPI. The next time around, after the 1997 energy agreement and renewed nuclear phase-out agenda, a similar industrial interest mobilization took place, but this time to no effect (see the second round in Chapter 5).

In subsequent processes, energy policy has been a relatively open arena. In principle, processes are influenced by competing interests such as industries, suppliers, distributors and users of different types as well as economic and environmental interest groups and academicians. Sweden is a small country and there are relatively clearly defined epistemic communities of researchers with rather predictable positions, both on the green side and on the industrial side (Nilsson, 2006). Still, for any specific policy preparation, there is a relatively strong variation in the representation of environmental interests in committees of inquiry. For instance, the Climate Commission set up to develop a comprehensive Swedish climate policy had an environmental representation that balanced the industrial side fairly well. On the other hand, the two committees of inquiry to examine the use of the flexible mechanisms of the Kyoto Protocol in Sweden – FlexMex 1 and 2 – largely excluded environmental interest groups and agencies (SOU, 2004:62). Their chairperson argued that environmental interest was inherent in the whole process, through the mandate and instructions given to the committee. And while environmentalists complained that they were not part of the process, these committees, as opposed to the more inclusive Climate Commission, finally came to agree on, and anchor with industry, very powerful and for some industries costly restrictions in the interest of protecting climate. The relationship between access to procedures and EPI is clearly a complex and ambiguous one. EPI depends on ownership and trust between the stakeholders involved in a process as much as making sure that environmental interests are properly represented.

Between 1997 and 2006, the Social Democrat government depended on the support of the Greens and the Socialists, both for political support in general and for environmental and climate policy in particular. What might be the effect of this position of the Greens? In general, they have induced political pressure on the government to implement EPI provisions and make policy bills greener than the Social Democrats probably would have preferred them to be otherwise. A concrete example is the political negotiations surrounding the climate policy bill in 2001, where the committee had reached a political agreement for a greenhouse gas reduction target but the Greens, in subsequent parliamentary negotiations, refused this and a new agreement was made for a doubling of the target (SOU, 2000:23; Prop, 2001/02:55). This was significantly more ambitious than the target allocated under the EU framework, which most other parties would probably have been happy with at the time. Similarly, the presentation of the climate bill (in 2006) was delayed several times because the Greens would not accept the (in their view) modest targets suggested by the government (Prop, 2005/06:172). It will be interesting to see how the shift in government in late 2006 will come to affect various green actors' access to policymaking processes.

Knowledge acquisition

Since it became a policy field of its own in the 1970s, energy policy has been subject to intense policy-analytical work, probably more so than other policy fields. Research studies, both commissioned and non-commissioned, have constantly fed the process. Large committees of inquiry have been popular within energy policy, such as the Climate Commission in 2000 or the Energy Commission of 1995. The very large committees have, in later years, been replaced with lower-key processes including ad hoc investigations, expert committees and evaluation programmes. This is part of a more general shift in Swedish policymaking towards smaller and quicker committees of inquiry (Riksrevisionen, 2004a). However, it also indicates a lower degree of political sensitivity of the energy issues.

One can also witness a more generic move towards more evidence-based policymaking – leaning more heavily on evaluations and institutionalizing them in the form of scheduled 'control-stations' (SEPA and STEM, 2004; Miljömålsrådet, 2002). This is to some extent the result of technical reporting requirements under international agreements, for instance relating to the Kyoto Protocol. At the onset of the new millennium, the government commissioned a great number of energy policy assessments to committees of inquiry as well as some external evaluations of the 1997 energy policy programme. Thus the government is at least paying significant lip service to evidence-based policymaking. But what has been the effect of these trends and shifts towards evidence gathering? The implications for EPI from this more 'technical treatment' of energy policy appear ambiguous. It is certainly true that a series of energy policy assessments between 1998 and 2001 fed into preparation processes of, for instance, the 2002 energy policy bill, the development of climate change policy and the handling of the nuclear phase-out programme, and it would be far from the truth to say that these assessments were subsequently ignored. Nonetheless, all energy policy bills rely on negotiated agreements with the cooperative parties (the Greens, Centre and Left). In such highly politicized processes, knowledge tends to be either ignored or only used for strategic purposes. For instance, the 1997 energy policy programme proposed that the closure of Barsebäck 2 be subject to conditions regarding environment and supply safety and needed to be subject to extensive impact assessments of electricity prices, investments, the environment, employment and redistribution, and the general functioning of the electricity market. Evaluations in the early 2000s showed that the conditions had not been fulfilled (COWI, 2000; ÅF Energikonsult, 2002). As a consequence, the MoI recommended that the conditions for closure of the second reactor in Barsebäck should be postponed but still be carried out before the end of 2003. A similar assessment was made one year later (Skr, 2000/01:15). However, again a parallel track of politics was pursued, with little apparent relation to the evaluations acquired. As opposed to party-political bargaining, the government this time tried to deal with industry directly and agree on a phase-out plan. However, the negotiations, which went on quietly for two years, were terminated towards the end of 2004, and on the same day as this was announced, the government announced the closure of Barsebäck 2. Hence the process of reaching political agreements moves forward – in parallel with, but disjointed from, knowledge acquisition. Knowledge is overlooked as soon as it does not conform with the political position. In some cases, even the actual gathering of evidence is avoided, as was the case during the presentation of the 1997 energy agreement, when the government openly rejected the parliamentary request to carry out an impact assessment of the proposed course of action (Virgin, 1998).

Still, new knowledge does contribute to shaping and reforming frames over time. A notable example is from the late 1980s, when energy systems analysts in Sweden presented a new frame of reference for thinking about energy policy and the energy sector. Instead of viewing the management of the sector as a technical optimization issue, they developed a socio-ecological risk perspective (Lönnroth et al, 1978). The view was that a 'solar society' was achievable, and that most barriers were institutional rather than technological. As we saw in Chapter 5, this had a strong imprint on the 1988 energy bill and its radical nuclear phase-out agenda.

Institutionalized EPI instruments

As described in Chapter 1, Sweden has actively pursued several institutional strategies for EPI, including the establishment of sector responsibility, environmental management systems, impact assessment procedures and national environmental quality objectives (NEQOs). How has this institutionalization played out in energy policymaking? In the long term, these procedures are expected to institutionalize and build up constituencies for certain types of knowledge and hopefully induce organizational learning. However, these processes are slow and difficult to discern and delineate empirically. In the short term, we have primarily witnessed that the implementation of sector responsibility without clear resource allocation led to compartmentalization of environmental issues. The typical picture, at both agency and ministry levels, has been that one junior desk officer is in charge of the environmental reporting and coordination of environmental issues, and that this is largely disconnected from the mainstream activities.

However, a more efficient, coordinated and, therefore, much more acceptable procedure is being established as a result of the NEQO system's more prominent role. This prominence is indicated, for instance, in the assignment of the main sector agency chiefs as members of its governing body, the NEQO Council. It is also seen in the connection of all other EPI systems to the NEQOs, such that the sector responsibility is focused on the NEQOs, the environmental management systems must report on the NEQOs and the impact assessment needs to address all NEQOs. With acceptance and ownership comes EPI. This process is, however, ongoing. The implications on energy policy formation from the NEQO system are still difficult to discern. As we have seen, in practical policy, the overriding concern is for climate change impacts, and it crowds out other issues. The systematic and comprehensive approach suggested by the NEQO system is therefore not realized in the overarching policy frame. Nonetheless, energy policymakers need to respond to and report on possible implications for each of the categories, for instance in the annual agency reporting and in the ministry-level impact assessment procedures. Whether these smaller procedures will have an impact on the real policy development remains to be seen.

Agriculture

In Chapter 5, we described how Swedish agricultural policy since the 1990s has evolved in three major policy rounds: first, the deregulation of agriculture in 1990, where subsidies to protect the rural landscape appeared for the first time (Ds, 1989:63; Prop, 1989/90:146); second, entering into the EU in 1995 and therefore the Common Agricultural Policy (CAP) and the establishment of the first environment and rural development programme (Prop, 1994/95:75); and third, the CAP reform beginning in 2003 (Ds, 2004:9). In the following sections, we discuss how institutional factors have influenced the extent of learning about environmental concerns within those developments.

Institutional and organizational landscape

The institutional landscape in agriculture has been stable over time at the national level. The National Board of Agriculture (NBA), under the Ministry of Agriculture, Food and Fisheries (MoA), has been the governmental agency in charge of agriculture and food policy since the post-war period. In 1991 the NBA went through reorganization at the regional level. Its previous regional bodies (*Lantbruksnämnder*) were established in 1948 with a mandate to render agriculture more efficient and became special agricultural units under the government's general County Administrations. This was meant to facilitate interaction and coordination with other units, such as planning, nature protection and the environment, cultural heritage, and regional development. Hence the administrative change followed from the deregulation of agriculture and was primarily related to the implementation of policy. It followed from an increased emphasis on environmental values at the time, rather than the other way around.

When Sweden joined the EU, the MoA and the NBA became responsible for implementing the CAP and the national Environment and Rural Development Plan (ERDP). This had an immediate effect on the organizations' workload, but the influence on value systems is not straightforward. The ministry largely maintains its professional identity from the days when production interests dominated the agricultural policy agenda. In practice, when the new agriculture-for-sustainability was to be implemented, our interviews show that two different groups formed within the ministry: one liaised with the Ministry of Finance, arguing for increased market orientation and deregulation, and another linked to environmental interests and was more concerned with sustainable development goals (Interview, SEPA 2, 2004). As we shall see below, this has resulted in two parallel 'tracks' in the agricultural policy debate, in which EPI is only visible in one. As expressed by an interviewee at the MoA:

> *If I think about the other committees [which are not specifically working with environment] then I must really ponder about how much [the environment is discussed] [...] In the committee for ecological farming of course environmental aspects are there, but that it is integrated in decision making, well, I cannot easily think of any occasion where it has been discussed on the agriculture side [...] as I said, we have really placed environment in the ERDP track. We solve it there, and then we can be rather hard on the rest. It is not 100 per cent integrated in that way.* (Interview, MoA 2, 2004)

Since the establishment of agricultural units at the County Administration, their work has become increasingly geared towards the ERDP, while the advisers from the Federation of Swedish Farmers (LRF) and Rural Economy and Agricultural Societies (REAS) are more production oriented. Nevertheless, the REAS was always targeted at small-scale farming, and prone to environmental thinking within its advice to farmers, while the LRF has recently changed from a 'production' to a new 'ecological farming' emphasis. It is now a member of the international ecological farming network, and its chairman is even an ecological farmer. Since 1992, the LRF's goal has been for Swedish farming to become 'the world's cleanest agriculture', carried forward in the 'Catch the Nutrients' campaign from 2001 (see below). The LRF has become very active in environmental issues since the change of leadership (Interview, SEPA 2, 2004). Hence both the LRF and the REAS have become important proponents for rural policy in a wide perspective, including all aspects of sustainable development, and contributed to EPI progress particularly in the second and third policy rounds. Again, however, this progress is most visible in the second policy 'track'.

Still, compared to the parallel forestry sector, which already promoted EPI in the 1980s and even changed its legislation towards balancing production and nature protection goals, it took much longer for agriculture to react to the increasing demands from environmental interests. Part of the slow progress can be attributed to difficulties within the NBA, in particular a low level of support to environmental goals from its leadership (Interview, SSNC 3, 2004). Even if officers at the county level had gradually changed their attitudes towards more emphasis on environmental concerns, they had little backing from the national agency. As the organization changed at the regional level, and the agriculture units became more integrated into other regional policymaking within the County Administrations, this contributed to a better understanding of environmental issues among those agricultural officers:

> *In some of the new agricultural units [Lantbruksnämnder, referring here to areas in southern Sweden] they have really integrated rural development, agricultural development, protection of nature and cultural heritage, tourism, etc in a very good way [...] it would not have happened without whole-hearted engagement from its staff. This is indeed a cultural revolution!* (Interview, SSNC 3, 2004)

As mentioned above, two very divergent tracks of agricultural policy seem to be developing in Sweden (and probably also elsewhere). The first stems from the period before EU membership, and relies on deregulation and open markets, in which commercial production of food on a large-scale basis is the main objective. The second builds on agriculture for preserving the cultural landscape and providing environmental services, and could possibly be regarded as a more conservative view of agriculture as a way to protect traditional values with small-scale production. The scale is important here: the role of agriculture in the global view differs largely from that of a national or local perspective, where the distributional issues and concerns for global justice are largely absent (Interviews, SSNC 2, 2004 and MoA 1, 2004). The second track also contains agriculture-for-development, which constitutes a main ingredient of the ERDP, although in Sweden it has been more focused on agriculture-for-sustainability than elsewhere in Europe. In 2002, however, the Swedish government took a special initiative to increase national coordination of regional policy development across 27 different government bodies. This initiative was headed by the Swedish Agency for Economic and Regional Growth (NUTEK) and concentrated on the regional growth programmes with a view to promoting sustainable economic growth throughout the country. Nevertheless, the regional development programmes were criticized for neither paying heed to environmental concerns (SEPA, 2001) nor becoming integrated with other ongoing sustainable development initiatives at the local level (Eckerberg and Dahlgren, 2005, p52). According to one of our informants, many County Administrations have also failed to integrate regional development with agricultural projects because they do not seem to apprehend the symbiotic relationship. Instead, the support for regional growth projects has been directed more towards industrial and technical projects (Interview, SEPA 2, 2004). Also within the government, there is little cooperation between the ministries of industry and agriculture, which contributes to the low priority given to agriculture in regional development policy (Interview, MoA 1, 2004).

In general, the political parties appear reluctant to take a clear stand on agricultural policies, possibly since public opinion is still hesitant regarding the virtues of Swedish EU membership and the CAP. It may also be questioned whether the Social Democratic Party in power is really keen to develop a profile on agriculture policy. The few farmers that remain constitute a very small proportion of the voters, and the countryside ceased to be an important political base some time ago, when urbanization brought the rural workers into towns and industries. The Centre Party, which had its traditional base in the rural areas, has diminished in size and new rural movements have become much more environmentally orientated than previous generations. Moreover, the consumer interests are rather weak in Sweden, compared to the UK, for instance, and bringing agriculture issues into politics would reveal their weakness in relation to the much stronger production side of agriculture. None of the larger political parties can risk such a debate.

Actors' access to policymaking

The agricultural sector is dominated by a few professional networks, and corporatist traditions have traditionally been intense. Since Sweden is a small country and agriculture is a comparatively small sector, collegial relations remain close and with frequent exchanges of staff between the major authorities and organizations. The LRF is a central actor, representing the interests of those who own and/or cultivate land and forests. It is also the basis for the agricultural-cooperative movement. There has been a remarkable growth in individual membership of the LRF in recent years, despite the decline in farming area. The LRF has switched from being a partner in negotiations with the state on the pricing of agricultural products (prior to 1990 when those negotiations were abolished with the new food policy) to becoming a modern lobby organization in national and EU policymaking (Eckerberg and Wide, 2001, p16). The relations between the farmers' organizations and public authorities remain strong (Micheletti, 1990; Rothstein, 1992). Some farmers also belong to the Rural Economy and Agricultural Societies (REAS) and the National Federation of Rural Community Centres. Both the LRF and REAS aim to promote the interests of farming and the rural community at the local level and offer expertise to advise their members. Farmers are thus able to obtain management advice both from their own interest organizations and from the state.

Experts and scientists have remained important in the development of policy. The agricultural sector shares a sector university with the forestry sector, the Swedish University of Agricultural Sciences (SUAS), in which professional norms and scientific research are produced. The Swedish Institute for Food and Agricultural Economics is another source of expertise. This enables an informal side to agricultural policy processes that may be of importance, although they are difficult to trace. The MoA formed an informal think-tank to advise policymakers in both EU negotiations and in Swedish implementation of the CAP. This group consists of selected experts from SUAS, the NBA, and the Swedish Institute for Food and Agricultural Economics in Lund who were also involved in committees of inquiry and other investigations over the years. This is an important forum for sharing ideas 'as one can speak freely, test different arguments, and discuss openly the various problems' (Interview, MoA 2, 2004). SUAS supplies both public and private agricultural organizations with research data and background studies. Its profile as an applied research institute, including training of agricultural expertise, has contributed to its considerable influence in the making of policy. For instance, SUAS provides environmental monitoring and state-of-the-environment reports for various sources of pollution and contributes with economic analyses of various policy options relevant to agriculture.

Consumer interests have, since 1992, been organized in an umbrella NGO entitled the Swedish Consumers' Association (Sveriges Konsumentråd), which speaks for 28 member organizations. Moreover, since 1994, the Swedish Consumer Coalition has organized some 16 more environmentally orientated NGOs.

These are relatively new movements, and compared to several other countries Swedish consumers are quite weakly organized. Moreover, they tend not to engage so much in agricultural policy, but rather in issues concerning the food industry as such (Interview, MoA 1, 2004). Possibly this has to do with the favourable image of Swedish farmers in the public debate, along with the lack of food-related scandals in Sweden as compared with other parts of Europe. In Sweden, the Minister of Agriculture is responsible for consumer policy also, and the Swedish Consumer Agency is its implementing agency in charge of looking after consumer interests. The Swedish KRAV, which represents organic farmers as well as other environmental interests, is the only acknowledged certifying organization in Sweden for green products and is a member of the International Federation of Organic Agriculture Movements. Even if consumers have the potential to 'vote with their feet' and are increasingly selecting green products (Micheletti, 2003), this pressure is not substantially guiding the policy agenda. A sign of this is the recent decline in number of ecological farms certified by KRAV (see also Chapter 1).

In recent years, organizations and interests outside agriculture (including environmentalists) have gained a stronger voice. The SSNC and the World Wide Fund for Nature (WWF) represent those environmental NGOs mostly engaged in agriculture policy (Eckerberg and Wide, 2001). The environmental movements have a large influence on media and public opinion, but a weak connection to specific decisions. Among environmental interest groups in Sweden, the SSNC has been much engaged in the discussions for the CAP reform, particularly in issues of biodiversity and the agricultural landscape, chemical residues from agriculture in water systems, GMOs, and the relation between EU agricultural policy and the third world. The WWF has a similar orientation, and has become known especially for its projects on maintaining biodiversity, including animal husbandry, according to traditional agricultural practices. Both the SSNC and the WWF have launched campaigns to protect the marine environment, with clear connections to agriculture, since much of the eutrophication stems from leaching out of farming areas. In addition to these two major environmental NGOs, the environmental association Friends of the Earth–Sweden has also been concerned with GMOs and agricultural trade, for example in the development of the CAP and World Trade Organization (WTO) rules. The environmental organizations have therefore played an important role in pressure for EPI.

In response to the most recent CAP reform, the debate in the interdepartmental working group centred on how the national envelope would be allocated – according to a farm-based or a regional model. The farm model implied that support would be based on the track record of production on each farm, whereas a regional model would be based on average production in a region. The Federation of Swedish Farmers argued strongly for the farm model, since it would entail less change for the individual farmer. For competition reasons, it also stated that the use of the national envelope should be very limited. In contrast, environmental interest groups (Interview, SSNC 2, 2004) argued for equal support across the country, regardless of previous production. In their view, while the reform was based on decoupling the support from production, it was only possible to justify society's costs of a maintained production with arguments about its contribution to a diverse landscape. In addition, the SSNC wanted to use the national envelope to subsidize the environmental goods associated with animal husbandry. As discussed in Chapter 5, the government settled for a compromise between the two models. This was based on calculations made both by the NBA (Ds, 2004:9) and alternative calculations made in the consultation procedure by the SSNC together with the Ecological Farmers (SSNC webpage, accessed 5 May 2006). In this policy round, the Green Party together with the SSNC and Ecological Farmers did attempt to influence, but their impact was only minor (Interview, SSNC 3, 2004).

Knowledge acquisition

As already emphasized, expert advice has an important influence on the preparation of agriculture policy, and this has also brought in new evidence on environmental impacts. In 1986, prior to the first policy round, SEPA had presented a report on agriculture and the environment that focused on three central problems generated by modern farming. These were nutrient leaching (with subsequent eutrophication of water and soils), the spreading of chemicals and heavy metals, and the loss of habitat and biodiversity (SEPA, 1986). The idea that agriculture alongside transportation and energy is one major source of environmental problems was further developed in the government bill 'Environmental policy for the 1990s' (Prop, 1987/88:85). The assumption that agriculture automatically entails resource stewardship and the integrity of the corporatist decision-making system within agriculture were questioned (Vail et al, 1994, p126). New ideas about agriculture also challenged traditional assumptions that agriculture should simply deliver foodstuffs. Environmental aspects were not only seen as externalities to the production of food, but public goods such as a diverse and open landscape, recycling of municipal waste, and production of biofuel to replace fossil fuels were now seen as complementary products of farms. These new insights were thus brought onto the political agenda and had a major influence on the design of new agri-environmental policy with the introduction of environmental fees on pesticides and chemical fertilizers in 1984 and grants for landscape preservation starting in 1986 (Eckerberg, 1994).

The Swedish tradition of creating working groups and investigative committees for policy development is notable in the agriculture sector. Several of these have contributed to including environmental instruments into the new policies, largely pressured from a combination of public opinion and scientific advice. Since the early 1990s, four major government committees of inquiry (SOU 1984:86; 1993:33, 1997:102 and 2002:75) and a long list of smaller expert-led committees (Ds Jo, 1987:2 and 1987:3; Ds Fi, 1988:54; Ds, 1989:50, 1991:87, 1998:70 and 2004:9; SOU 1992:14, 1995:88, 1997:25, 1997:150, 1997:167, 2001:68 and 2003:105) were set up for advising on agricultural policy. Some of the expert-led ones have specifically dealt with consumer interests (SOU, 1994:119 and 1996:62), organizational aspects (SOU, 1990:87, 1996:65 and 1998:147), issues of finance (SOU, 1995:117 and 1998:108) and GMOs (DsJo, 1990:9; SOU, 1992:82; Ds, 1996:73). Some of the expert-led committees involved reference groups (including representatives from political parties) that could influence the development of the proposals. However, the investigations have neither been publicly debated to any significant extent (at least not compared with energy issues), nor become much politicized.

Over time, the availability of data on environmental aspects of agriculture has improved considerably. For example, the database on threatened species at the Swedish Species Information Centre was established in 1991 at SUAS, in collaboration with SEPA and the WWF. Apart from monitoring information on rare and threatened species, this centre contributes to action and management plans. Impacts from agriculture, along with forestry, are highly recognized in the monitoring, and the quality of the information has improved substantially over the last decade (Interview, SSNC 3, 2004). There is, according to our interviewee from the LRF, a need to collect information from different angles and critically assess it, but he believes that this process has improved over time (Interview, LRF 1, 2004).

One interviewee from the MoA emphasizes that EPI in Sweden works well in comparison with the situation in the EU, where analyses of environmental effects are seldom carried out even when major decisions are taken (Interview, MoA 1, 2004). Within the EU, the Commission is responsible for assessing the environmental effects of the CAP and other rural programmes, but in practice such information is largely absent. For instance, in the Commission's proposal on the mid-term review of the CAP, the emphasis was on the economic effects and the environment was hardly mentioned (Interview, MoA 1, 2004). Environmental issues within agricultural policy are only discussed at the EU level in the committee for ecological farming and cannot be traced in other committee work (Interview MoA 2, 2004).

Another interviewee from SEPA confirms that, in Sweden, the basis of knowledge is quite good within agriculture compared with many other sectors. There are a number of evaluation studies and analyses of consequences, such as the studies published by SEPA under the umbrella of 'Environmental Sweden in 2021' (Interview, SEPA 2, 2004; SEPA, 1999). In this future study, the effects on sustainable development, including in agriculture, were assessed through different scenarios at both the global and national levels, though these might still be difficult to translate into concrete decisions (Interview, SEPA 2, 2004). Nevertheless, when the CAP revisions were to be formulated into national policy in the third policy round, the SSNC, together with ecological farmers, the consumer movement (Sveriges Konsumentråd) and the trade union for food workers, made an input to this process. They demanded an assessment of the consequences for both the environment and consumers in the material provided by the government since such information was absent (Interview, SSNC 2, 2004).

The time factor is considered to be a general problem for systematic analysis of various consequences of the CAP in regions with different natural and socio-economic characteristics. Since the NBA is assigned to assess the national consequences of EU policy, and its traditional focus has largely been placed on socio-economic aspects of agriculture, environmental aspects might easily be given less attention and there is insufficient time for independent review from various interest organizations (Interview, SEPA 2, 2004). Furthermore, it is common to only focus on certain details and thereby lose a holistic perspective. The limited time allocated for discussing potential policy impacts also has a bearing on the legitimization and subsequent implementation phase (Interviews, SSNC 2, 2004 and SEPA 2, 2004). Moreover, to allow for practical use of the assessments throughout the policy process, several of the respondents emphasized that it is important that all types of relevant information is presented in the same document, but that it should be clarified for each aspect. Thus the balancing between interests can become more visible, and politicians may adjust the priorities along the way.

Institutionalized EPI instruments

In 1998 the government allocated sector responsibility for implementing environmental goals within agricultural policy to the Ministry of Agriculture and the NBA, along with a requirement to introduce environmental management systems. Like in the other state organizations, the EMS has dealt with internal operation of activities rather than their content. Environmental reporting applies to the gathering of statistics, but we see no evidence that this information is used proactively in the preparation of policy, nor in systematic *ex post evaluation*. As for the energy sector, an explanation might be that only a few officers are assigned the tasks of environmental reporting and the systems thus tend to be compartmentalized. The information can, however, be brought onto the agenda by experts and environmental interest groups, as we have seen above.

The process surrounding the NEQOs constitutes a platform for potential EPI, and probably more so than in the energy sector. Agricultural practices are pivotal to many of the NEQOs, especially those to maintain a varied agricultural landscape and to protect biodiversity. However, apart from its apparent impacts on landscapes, wetlands and biodiversity, agriculture also contributes to all sorts of pollution as shown in Chapter 4. Many of these NEQOs are indeed discussed and targeted through various measures within agriculture, especially as they relate to land use practices. But even though agriculture, together with food consumption, uses a great deal of energy and contributes significantly to CO_2 emissions, neither transport issues nor energy policy goals have been brought to attention within the agriculture sector (Interview, MoA 1, 2004). Nevertheless, our interviewee from LRF emphasized the global ecological footprints of food consumption from far distances, in which Swedish farmers are viewed as much more environmentally friendly both in their production techniques, such as use of water and pesticides, and in the transport of goods to Swedish consumers. There is, apparently, a discrepancy between the national and global perspectives of Swedish agriculture with relation to energy consumption.

The increased emphasis on environmental concerns in the second policy round has led to a collaborative project based on the NEQOs. This project, entitled 'Catch the Nutrients' (which in Swedish – *Fånga Näringen* – has two connotations, since it can also mean 'grasp the [agricultural] business'), was initiated in 2001 between the agricultural organizations (including the LRF) and the state authorities (the NBA and the County Administrations). It includes an information campaign to raise awareness among farmers and create environmental plans at the farm level for improved management, in particular related to eutrophication. The project began in southern Sweden and has since 2005 extended to the middle part of the country, thus covering most of Sweden's agricultural land. Special advisers have been designated, and farmers can also access information on the internet. At present about 6000 members have signed up (www.greppa.nu/, accessed 30 January 2006). At the local level, and for certain environmental issues, EPI thus seems to have taken place.

Bioenergy

As mentioned briefly in Chapter 1, the most prominent institutional feature distinguishing bioenergy from the other two policy areas studied here is its multisectoral characteristics, and specifically that it is dependant on both energy and agricultural policy. Another important factor in bioenergy policy, not to be neglected, is the need for economic 'carrots' in order to induce farmers to alter their production from traditional (food) to new products (such as energy crops) (Interviews, SSNC 3, LRF 2 and STEM 6, 2004). Thus coordinated stimulation of both the demand side and the supply side of biomass production are essential ingredients of a sustainable bioenergy policy.

This makes the successful achievement of environmental policy integration in bioenergy policy dependent on three factors. First, vertical EPI in the two affecting sectors is necessary separately. Second, it requires that bioenergy is propounded based on environmental arguments as a result of the vertical EPI. These two factors have been discussed in Chapter 5. And third, horizontal policy coordination of energy and agricultural bioenergy policy substantially increases the likeliness for EPI to materialize in policy practice; this will be further elaborated below.

Chapter 5 presented three changes in bioenergy policy frames. In the early 1990s there was a boost in bioenergy research and production materialized, partly as a result of the incorporation of bioenergy-for-sustainability and bioenergy-for-agricultural-adaptation into bioenergy policy. In the mid 1990s the sustainability emphasis grew stronger but, paradoxically, this occurred simultaneously with drastically lowered subsidies for energy forest cultivation. In the early 2000s a move towards accepting energy sources such as natural gas and waste occurred in Swedish energy policy, together with a frame shift towards more market measures in bioenergy policy. These frame changes constitute a point of departure for the institutional analysis.

Institutional and organizational landscape

As bioenergy can be described as floating between two sectors, its main institutional host has varied over time, as has the degree of horizontal policy coordination. The MoI undertook the role as a driver during the 1980s, but the interest in agricultural biomass production grew in the late 1980s and early 1990s in step with a decreasing security focus in Sweden, leaving agriculture devoid of its previous economic support. With both energy and agricultural policy set on increasing the supply and demand for biomass for energy production, bioenergy policy became more coordinated than ever. Energy forest cultivation, for instance, increased rapidly as a result of massive economic subsidies based on LRF and MoA arguments for bioenergy-for-agricultural-adaptation. The LRF has since then been one of the most prominent proponents of agricultural biomass production, and has over time moved towards incorporating bioenergy-for-sustainability as a strong argument for agroenergy (energy produced in agriculture) development. In the 1991 bill on energy policy – a proposal presented by the MoI and the MoA in concert – the powerful measures to stimulate biofuel-based power and heat production was partly motivated by agricultural policy. The introduction of financial support for energy crop cultivation was expected to lead to a rapid increase in the supply of biofuels and the MoI argued that it was important to promote the use of biofuels in power and heat production as part of the development of a sustainable bioenergy market (Prop, 1990/91:88). During the first policy round, vertical EPI in the two sectors thus resulted in sustainable outcomes due to a high degree of policy coordination in the area of bioenergy.

At the same time as the Swedish government proclaimed the goal to create a Swedish green welfare state during the mid 1990s, bioenergy moved towards 'bioenergy-as-entrepreneurial-practice', leaving main developmental responsibility to individual farmers and energy entrepreneurs in industry. This occurred during the same period of time as the Swedish implementation of CAP, restricting national governmental ability to provide farmers with economic support for energy crop cultivation. As a result, bioenergy once again developed into an energy sector issue – leading to more emphasis on demand-stimulating, market-oriented measures than biomass supply stimulation. Sweden lobbied for EU subsidies for alternative agricultural production, but only slight increases in EU subsidies for energy forest cultivation have occurred during the 1990s and early 2000s. This may provide an explanation for an increasing interest in other energy sources in Sweden at the cost of agricultural biomass production. Coordination was spurred and agroenergy was interesting for Sweden only as long as national policy coordination could lead to practical results. When the power over agricultural policy moved to the EU level, interest was relocated to pure energy sector areas where national influence was intact. Even though energy policy goals were recognized in agricultural policy, cooperation between the MoI and MoA was lacking and became an obstacle to Swedish agroenergy development (Interviews, MoA 1 and STEM 6, 2004). Given this, it is not surprising that agricultural bioenergy production has been fallowed during the last decade. During the last ten years, lack of policy coordination between CAP and Swedish energy policy, and less cooperation between MoI and MoA have been obstacles to Swedish agroenergy development.

In the mid 2000s, however, the Russia–Ukraine energy conflict induced a renewed interest in a European energy policy, including support for domestic energy that can be obtained from forestry, waste materials and agriculture (EC, 2005a; EC, 2005b). Circumstances similar to those in Sweden during the early 1990s seem to have boosted bioenergy policy coordination at the EU level. The desire for a secure European energy supply, together with calls for reformed agricultural production, created the conditions for an integrated energy policy to form, including actors in forestry, agriculture and waste as important energy producers. In Sweden the establishment of the Ministry for Sustainable Development, assembling environmental and energy issues under the same roof, led to a revival for agricultural biomass production. However, the Ministry for Sustainable Development became short-lived. In 2006, the new Liberal–Right government shifted back to a more conventional model of a Ministry for the Environment. The new Minister for

Sustainable Development has argued in favour of energy crops on a number of occasions (Sahlin, 2005). It remains to be seen how these recent institutional shifts will affect policy coordination in bioenergy policy in Sweden and in Europe.

Actors' access to policymaking

According to our interviewee at the Swedish Energy Agency, the network of actors active within the area of bioenergy policy in Sweden is relatively large, but quite open. Many of the actors know each other, and informal contacts are common (Interview, STEM 5, 2005). With regards to agricultural biomass production, the LRF and the Swedish Bioenergy Association (Svebio), together with the SSNC, the Swedish Farmers Supply and Crop Marketing Association and the Swedish Agricultural Workers' Union (SLF), stand out as the most enthusiastic proponents throughout the studied period (Interview, STEM 6, 2004). Already in the debate preceding the 1988 decision on nuclear phase-out, the LRF and SveBio started to call attention to the environmental benefits of replacing nuclear and fossil energy with domestically produced bioenergy; and they continued to do so in the debate on the new food policy as well as the new energy policy (Prop, 1986/87:159). SUAS has also been a prominent pro-bioenergy actor in the policy process, showing that expert input is considered important in bioenergy policymaking.

However, critical voices have also been raised. For example, the Swedish Forest Industries Federation (SFIF), in line with other industrial interests, were strongly against the idea of making nuclear phase-out dependent on an 'unrealistic' reliance on bioenergy, since they regarded the supply of wood fuel as insufficient and feared a future scarce supply of wood and energy. Concerning energy forest, the SFIF stressed the fact that such production was in the experiment stage and would probably not contribute to the Swedish wood fuel supply 'within a foreseeable future' (Prop, 1986/87:80). Fear of competition with energy companies for wood as raw material, and competition from agricultural wood supply, are possible explanations for the resitance from the pulp and paper industry towards bioenergy and energy forest. Power and heat production companies displayed doubts about energy forest as fuel at the start, but this changed rapidly. Oil companies and the car industry have traditionally been strong opponents against bioenergy in general, but in the 2000s a change in attitudes is visible (Interview, LRF 2, 2004).

Although the agricultural biomass lobby was successful during the agricultural deregulation era in the early 1990s, their influence over Swedish agricultural policy has drastically decreased during the last ten years due to the shift in policy levels, from the national to the supranational (EU) level. Instead, bioenergy proponents are now active in Swedish energy policy, where the competition for influence (compared with the agricultural sector) is more diversified. Apart from competing with strong actors (industrial spokesmen, for example) opposing bioenergy expansion, bioenergy proponents are up against other renewable energy sources such as solar energy, geothermic energy, water power and wind power. For example, a broad range of representatives were involved in the electricity certificate investigation of 2000–2001, most of whom represented divergent branches of the energy sector. However, the MoE, Ministry of Finance (MoF), SFIF and SEPA were also engaged in the investigation as experts (SOU, 2001:77).

The investigation also consulted interest organizations such as Svebio, Nordic Windpower AB and the Swedish Peatfuel Association. The car industry expressed doubts regarding biofuels in response to the 1996 renewable fuel investigation (SOU, 1996:184; Prop, 1997/98:145). In the 2004 renewable fuel investigation the special investigator met with a reference group representing the SFIF, oil companies and pro-car lobbyists as well as Greenpeace, the SSNC, the BioAlcohol Fuel Foundation (BAFF), Svebio and the LRF (SOU, 2004:4).

Thus, since many strong actors are active within energy policy in general, competition for influence is tough. Bioenergy proponents have had to coordinate their actions in order to gain influence. Svebio, founded in 1980, organizes over 400 members, of which 300 are bioenergy companies, including the LRF, SUAS, Swedish District Heating and the Swedish Farmers Supply and Crop Marketing Association Energy Section (Svebio, 2006), and has been an important Swedish bioenergy actor for over 20 years. The shift in the mid 1990s regarding which actors have a role in the formulation of bioenergy policy clearly mirrors the policy frame changes in the last decade.

Knowledge acquisition

Knowledge acquisition, monitoring and review have been important ingredients of the Swedish bioenergy policy and have been included in the discussions of many energy and agricultural committees over the years. In particular, the 1991 energy agreement obliged the government to evaluate and report on results from the measures to promote energy conservation and new power and heat production, as well as to report on Swedish nuclear safety and suggest further measures annually in the budget bill (Prop, 1990/91:88).

Three governmental investigations with a specific bioenergy focus have been produced during the studied period of time: two MoE renewable fuel special investigator studies (SOU, 1996:184 and 2004:4) and one MoE/MoA parliamentary 'biofuel commission'. The biofuel commission had an explicit coordinative mission and their work (SOU, 1991:93 and 1992:90) led to a bill on the promotion of biofuel use (Prop, 1991/92:97). In 1993 an interdepartmental working group, including the MoF, MoA, MoI and MoE, was appointed to suggest measures to 'remove possible barriers to an economically defendable expansion of combined heat and power (CHP) and biofuel use' (SOU, 1995:139, p357). Despite this policy coordinative measure, however, the group did not manage to reach a joint conclusion. Two different proposals were presented (Ds, 1994:28), neither of which was implemented.

The 1997 energy policy bill enjoined STEM to develop methods to measure the efficiency and the environmental, technological and economical impacts of the energy policy programmes. When designing the subsidies for biofuel-based CHP, the pros and cons of the previous subsidy system were to be taken into account in order to ensure the greatest efficiency possible (Prop, 1996/97:84). However, the bills on sustainable energy and sustainable fisheries/agriculture should be seen in the light of the broad-based governmental concentration on sustainable development (Skr, 1997/98:13; Prop, 1997/98:145). Evaluations were made of the measures for sustainable development undertaken in the different sectors. For example, the results from the economic support for cultivation of energy forest on agricultural land were evaluated on a yearly basis between 1999 and 2002 (Skr, 1998/99:5, 1999/2000:13, 2000/01:38 and 2001/02:50). Measures for sustainable agriculture were also evaluated by the LRF (LRF and SCB, 2001). It is possible that the measures undertaken would have increased environmental learning and translated rhetoric into action under different circumstances. However, the gap between the CAP and Swedish energy policy impeded such action.

Investigations of environmental consequences and cost–benefit analyses regarding bioenergy have thus been important parts of the policy formulating process. For example, the 1996 renewable fuels investigation was preceded by nine expert reports covering, for example, life-cycle analysis and environmental evaluations of emissions from and comparisons between different kinds of motor fuels (SOU, 1996:184). In the final report, the suggestions drew heavily from the results from the studies undertaken (SOU, 1996:184). The 2004 renewable fuels investigation was not granted similar resources. Nevertheless, consequence analyses, life-cycle analyses and supply potential of different biofuels were provided. The SOU (2004:133) concluded, for example, that cellulose biomass from forestry and agriculture had greater potential than cereal biomass to provide low-cost biofuels in the future.

Thus environmental input into the early stages of the policy process has been provided in bioenergy policy. However, the outcome of the process is dependent on negotiations between political parties and thus learning in policy practice depends heavily on which actors are involved in policy negotiations.

Institutionalized EPI instruments

NEQOs do not constitute a major part of bioenergy policy, even though bioenergy policy is evaluated with respect to these. According to the SOU (2001:77), it is 'difficult to find any environmental quality objective which is entirely unaffected by the production of electricity from renewable energy sources'. For bioenergy in particular, four environmental quality objectives are especially affected: reduced climate impact, clean air, natural acidification only and zero eutrophication. A shift towards increased utilization of bioenergy might cause increased emissions of particles, but it would still help to reach the goals of these four NEQOs. CO_2 emissions are expected to decrease, while emissions of SO_2, NO_x and VOCs (volatile organic compounds) can be expected to increase. To what extent, however, is a question of choice of technology and type of fuel (SOU, 2001:77).

More important instruments for bioenergy development are the specific goals for renewable energy included, for example, in the 1997 and 2002 energy agreements (Prop, 1996/97:84; Prop, 2001/02:143). In 1997 electricity production from bioenergy was set to increase by 0.75TWh per year. In 2002 renewable energy was set to increase by 10TWh from 2002 to 2010 through the new electricity certificates whereby different renewable energy sources compete with each other on the market. In order to follow up the results, the government proposed to use indicators on, for example, the possibilities to reach the 10TWh increase goal, the effects from the certificate system on renewable energy sources and commercial investments in bioenergy (Prop, 2001/02:143).

In general, EPI has occurred over time in bioenergy policy. Today, environmental consequences from bioenergy production and use are analysed and taken into account to a much larger extent than 30 years ago. Even though increased European influence put a stop to the highly coordinated and progressive Swedish bioenergy policy in the mid 1990s, institutional measures during the early 2000s may indicate a shift towards a more integrated bioenergy policy in Sweden and in Europe.

Comparison and Conclusions

The institutional analysis shows that corporate traditions between the state organizations and major economic interest groups remain strong. These power relations have, to a substantial extent, restrained environmental interest groups from obtaining access to policymaking processes, even though access has increased somewhat over time. Both sectors were reorganized during the studied time period, but this organizational change seems to have had only marginal influence on how policy has been formulated with regard to environmental goals, albeit perhaps allowing more interaction between various interests within the implementation of policy.

There are also some clear differences between the two sectors of energy and agriculture (see Table 6.1). First, energy appears much higher on the agenda of the

Table 6.1 *Comparison of characteristic features over time for the three sectors*

	Energy	Agriculture	Bioenergy
Institutional and organizational landscape	Corporatist network quickly eroding, many institutional changes towards market-orientation	Corporatist culture gradually eroding, two different policy strands emerging	'Floating' between two policy sectors, where agricultural side is very EU-dependent
Political attention	Since the 1970s very high political attention except in recent periods	Continuously low attention by political parties	On average medium attention as part of agricultural/ecological transition
Knowledge acquisition	More and more 'evidence-based' policymaking, traditionally heavy on expert assessments	National environmental data-bases have improved and are increasingly used in national assessment of policy	Research important policy instrument, experts often involved in committee work
Access of actors	Industrial actors traditionally prioritized but gradual increase in environmental access	Agricultural interests remain influential while environmental interests increasing	Agricultural bioenergy proponents less influential due to increased energy sector focus
EPI Systems	NEQO system plays a minor role so far, only really dealing with climate goal (which is treated separately)	Large number of NEQOs applicable, but remain environmental goals rather than agricultural	Many NEQOs affected by bioenergy, but not a major part of bioenergy policy; renewable energy goals more important

Swedish political parties. In particular, the nuclear phase-out agenda has drawn much attention in the public debate and led to a number of negotiations between political parties. By contrast, the agriculture sector and the bioenergy issue provoke little political debate. Farmers are still viewed by the public as environmentally benign, and no one argues against bioenergy production in principle. This might be explained by the political parties' reluctance to discuss the overall policy goals of agriculture, such as whether Swedish agricultural land should be mainly used to produce food or environmental services in the form of open landscapes, biodiversity or bioenergy.

Both sectors have acquired knowledge from various government committees and evaluations, which also has fed into the policymaking processes. Expert knowledge has played an important role in this respect, and is particularly notable within agricultural policy. It is difficult to evaluate, however, which types of knowledge have been most influential in the end, and to what extent, since decision-making processes are not always transparent. In the energy policy field, there has been much political bargaining – even outside of the sector – which leads us to conclude that the opening of the sector to environmental concerns was a necessary, but still not sufficient, condition for EPI.

There have been signs of a gradual change in values within the organizations involved in policymaking. Environmental aspects have emerged as part of the sectors' own definitions of their mandates, and integration of those aspects has thereby increased over time. However, despite this increased environmental awareness among the actors, those environmental integration systems that are put in place by the government (sector responsibility, EMS, NEQOs and other environmental reporting systems) seem inadequate for EPI to happen regularly. In particular, lack of time is highlighted as a constraint to assessing the environmental effects of policy options. This has also become an increasing problem with the increasing demands relating to EU policymaking procedures. In addition, we have noted that environmental reporting within the sectors' organizations has been allocated only to a few individuals, which compartmentalizes EPI. As shown in the analysis of agriculture, this sector can even be divided into two different strands, in which only that connected with the ERDP allows for environmental policy goals to become integrated into policymaking.

Chapter 7 provides a further and extended discussion and comparative analysis of the institutional aspects of EPI. Chapter 8 gives a brief account of some key lessons learned.

References

ÅF Energikonsult (2002) 'Barsebäck 2: Underlag för prövning av stängning' ['Barsebäck 2: Factual basis for consideration of closure'], unpublished material available from Regeringskansliet, Stockholm

Anshelm, J. (2000) *Mellan frälsning och domedag: Om kärnkraftens politiska idéhistoria i Sverige 1945–1999 [Between Salvation and Armageddon: On Nuclear Political History in Sweden 1945–1999]*, Brutus Östlings Bokförlag, Stockholm

COWI (2000) 'Evaluering af 1997s energipolitiske program: Analyse af udvalgte aspekter' ['Evaluation of the 1997 energy policy program: Analysis of selected aspects'], unpublished material available from Regeringskansliet, Stockholm

Ds (1989:50) 'Jordbruket och miljön' ['Agriculture and environment'], Environmental Advisory Council, Regeringskansliet, Stockholm

Ds (1989:63) 'En ny livsmedelspolitik' ['New food policy'], Ministry Publication Series, Regeringskansliet, Stockholm

Ds (1991:87) 'Arbetsgrupp för naturvårdshänsyn och de areella näringarna' ['Working group for nature protection and agriculture and forestry'], Interdepartmental Group of the Ministry of Environment, Ministry Publication Series, Regeringskansliet, Stockholm

Ds (1994:28) 'Förändrad kraftvärmebeskattning: Rapport från arbetsgruppen om kraftvärmebeskattning' ['A changed power-heat taxation: Report from the working-group on power-heat taxation'], Ministry Publications Series, Regeringskansliet, Stockholm

Ds (1996:73) 'Biodiversitet och framtida genpolitik' ['Biodiversity and future GMO policy'], Ministry Publication Series, Regeringskansliet, Stockholm

Ds (1998:70) 'Agenda 2000 – With reference to Swedish agriculture', Ewa Rabinowics et al (eds), Ministry Publication Series, Regeringskansliet, Stockholm

Ds (2004:9) 'Genomförandet av EU:s jordbruksreform i Sverige' ['Implementation of reformed CAP in Sweden'], Interdepartmental Group of the Ministry of Agriculture, Ministry Publication Series, Regeringskansliet, Stockholm

DsFi (1988:54) 'Alternativ i jordbrukspolitiken' ['Alternatives in agricultural policy'], Ministry Publication Series, Regeringskansliet, Stockholm

DsJo (1987:2) 'Åtgärder för att minska spannmålsöverskottet och stimulera alternativ markanvändning' ['Measures to reduce agricultural surplus and stimulate alternative land use'], Ministry Publication Series, Regeringskansliet, Stockholm

DsJo (1987:3) 'Intensititen i jordbruksproduktion' ['Intensity in agricultural production'], Ministry Publication Series, Regeringskansliet, Stockholm

DsJo (1990:9) 'Genteknik – Växter och djur' ['Gene technology in animals and plants'], Ministry Publication Series, Regeringskansliet, Stockholm

EC (2005a) *EU biomass action plan*, COM 2005:628, European Commission, Brussels

EC (2005b) 'Renewable energy: European Commission proposes ambitious biomass and biofuels action plan and calls on Member States to do more for green electricity', IP/05/1576 press release, Brussels, 7 December 2005

Eckerberg, K (1994) 'Consensus, conflict or compromise? The Swedish case' in K. Eckerberg, P. K. Mydske, A. Niemi-Iilahti and K. Hilmer-Pedersen (eds) *Comparing Nordic and Baltic Countries – Environmental Problems and Policies in Agriculture and Forestry*, TemaNord 1994:572, Nordic Council of Ministers, Copenhagen

Eckerberg, K. and Dahlgren, K (2005) 'LIP in Central Government perspective', in K. Eckerberg (ed) *Understanding LIP in Context*, Report 5454, Naturvårdsverkets förlag, Stockholm

Eckerberg, K. and Wide, J (2001) 'The nature of rural development', Report 2001:1, Department of Political Science, Umeå University, Umeå

Kaijser, A. (2001) 'From tile stoves to nuclear plants – The history of Swedish energy systems', in S. Silveira (ed) *Building Sustainable Energy Systems: Swedish Experiences*, AB Svensk Byggtjänst and Swedish National Energy Administration, Stockholm

LRF and SCB (2001) *Miljöredovisning för svenskt jordbruk 2000 [Environmental Reporting for Swedish Agriculture 2000]*, Lantbrukarnas Riksförbund, LRF, Stockholm

Lönnroth, M., Johansson, T. B. and Steen, P. (1978) *Sol eller uran: Att välja energiframtid [Sun or Uranium: Choosing an Energy Future]*, Liber, Stockholm

Micheletti, M. (1990) *The Swedish Farmers' Movement and Government Agricultural Policy*, Praeger, New York

Micheletti, M. (2003) *Political Virtue and Shopping: Individuals, Consumerism, and Collective Action*, Palgrave, Macmillan, Basingstoke, UK

Miljömålsrådet (2002) *Miljömålen – Når vi fram? De Facto 2002 [The Environmental Objectives – Will we Get There?]*, Naturvårdsverkets förlag, Stockholm

Nilsson, M. (2006) 'The role of assessments and institutions for policy learning: A study on Swedish climate and nuclear policy formation', *Policy Sciences*, vol 38, pp225–249

Prop (1986/87:80) 'Om forskning' ['On research'], Government Bill, Regeringskansliet, Stockholm

Prop (1986/87:159) 'Om vissa utgångspunkter för energisystemets omställning' ['On certain points of departure for energy system restructure'], Government Bill, Regeringskansliet, Stockholm

Prop (1987/88:90) 'Om energipolitik inför 1990-talet' ['On energy policy for the 1990s'], Government Bill, Regeringskansliet, Stockholm

Prop (1989/90:146) 'En ny jordbrukspolitik' ['New agriculture/food policy'], Government Bill, Regeringskansliet, Stockholm

Prop (1990/91:88) 'Om energipolitiken' ['On energy policy'], Government Bill, Regeringskansliet, Stockholm

Prop (1991/92:97) 'Om främjande av biobränsleanvändningen' ['On promotion of biofuel use'], Government Bill, Regeringskansliet, Stockholm

Prop (1994/95:75) 'Vissa livsmedelspolitiska åtgärder vid ett medlemskap i Europeiska unionen' ['Certain food policy measures with EU membership'], Government Bill, Regeringskansliet, Stockholm

Prop (1996/97:84) 'En uthållig energiförsörjning' ['A sustainable energy supply'], Government Bill, Regeringskansliet, Stockholm

Prop (1997/98:2) 'Hållbart fiske och jordbruk' ['Sustainable fisheries and agriculture'], Government Bill, Regeringskansliet, Stockholm

Prop (1997/98:145) 'Svenska miljömål. Miljöpolitik för ett hållbart Sverige' ['Swedish environmental objectives. Environmental policy for a sustainable Sweden'], Government Bill, Regeringskansliet, Stockholm

Prop (2001/02:55) 'Sveriges klimatstrategi' ['Sweden's climate strategy'], Government Bill, Regeringskansliet, Stockholm

Prop (2001/02:143) 'Samverkan för en trygg, effektiv och miljövänlig energiförsörjning' ['Cooperation for a safe, efficient and environmentally friendly energy supply'], Government Bill, Regeringskansliet, Stockholm

Prop (2002/03:40) 'Elcertifikat för att främja förnybara energikällor' ['Electricity certificates to promote renewable energy sources'], Government Bill, Regeringskansliet, Stockholm

Prop (2003/04:132) 'Handel med utsläppsrätter' ['Trade with emissions rights'], Government Bill, Regeringskansliet, Stockholm

Prop (2005/06:172) 'Nationell klimatpolitik i global samverkan' ['National climate policy in global cooperation'], Government Bill, Regeringskansliet, Stockholm

Radetzki, M. (2004) *Svensk energipolitik under tre decennier [Swedish Energy Policy During Three Decades]*, SNS Förlag, Stockholm

Riksrevisionen (2004a) *Förändringar inom kommittéväsendet [Changes in the Committee System]*, Riksrevisionen, Stockholm

Riksrevisionen (2004b) *Vattenfall AB – Uppdrag och statens styrning [Vattenfall AB – Mission and Governmental Control]*, Riksrevisionen, Stockholm

Rothstein, B. (1992) *Den korporatistiska staten: Intresseorganisationer och statsförvaltning i svensk politik [The Corporatist State: Interest Organizations and Public Administration in Swedish Politics]*, Norstedt Juridik, Stockholm

Sahlin, M. (2005) 'DN Debatt: Enklare lagar påskyndar byggandet av vindkraftverk', *Dagens Nyheter*, 13 March, p6

SEPA (1986) *Jordbruket och miljön: Handlingsprogram [Agriculture and Environment: Action Plan]*, Naturvårdsverket, Stockholm

SEPA and STEM (2004) *Sveriges klimatstrategi – Ett underlag till utvärderingen av det svenska klimatarbetet [Sweden's Climate Strategy – A Basis for the Evaluation of the Swedish Climate Efforts]*, Naturvårdsverket and Energimyndigheten, Stockholm

Skr (1997/98:13) 'Ekologisk hållbarhet' ['Ecological sustainability'], Government communication, Regeringskansliet, Stockholm

Skr (1998/99:5) 'Hållbara Sverige – uppföljning och fortsatta åtgärder för en ekologiskt hållbar utveckling' ['Sustainable Sweden – Evaluation and further measures to support an ecologically sustainable development'], Government communication, Regeringskansliet, Stockholm

Skr (1999/2000:13) 'Hållbara Sverige – Uppföljning av åtgärder för en ekologiskt hållbar utveckling' ['Sustainable Sweden – Evaluation of measures to support an ecologically sustainable development'] Government communication, Regeringskansliet, Stockholm

Skr (2000/01:15) 'Den fortsatta omställningen av energisystemet m.m.' ['The continued transformation of the energy system etc'], Government communication, Regeringskansliet, Stockholm

Skr (2000/01:38) 'Hållbara Sverige – Uppföljning av åtgärder för en ekologiskt hållbar utveckling' ['Sustainable Sweden – Evaluation of measures to support an ecologically sustainable development'], Government communication, Regeringskansliet, Stockholm

Skr (2001/02:50) 'Hållbara Sverige – Uppföljning av åtgärder för en ekologiskt hållbar utveckling' ['Sustainable Sweden – Evaluation of measures to support an ecologically sustainable development'], Government communication, Regeringskansliet, Stockholm

SOU (1984:86) 'Jordbruks- och livsmedelspolitik' ['Agriculture and food policy'], main report of the Food Policy Committee of 1983, Government Committee Report, Regeringskansliet, Stockholm

SOU (1990:87) 'Den nya centrala jordbruksmyndigheten' ['The new central agricultural administration'], Government Committee Report, Regeringskansliet, Stockholm

SOU (1991:93) 'El från biobränslen' ['Electricity from biofuels'], Government Committee Report, Regeringskansliet, Stockholm

SOU (1992:14) 'Mindre kadmium i handelsgödsel' ['Less cadmium in chemical fertilizers'], Government Committee Report, Regeringskansliet, Stockholm

SOU (1992:82) 'Genteknik en utmaning' ['Gene technology: A challenge'], Government Committee Report, Regeringskansliet, Stockholm

SOU (1992:90) 'Biobränslen för framtiden' ['Biofuels for the future'], Government Committee Report, Regeringskansliet, Stockholm

SOU (1993:33) 'Åtgärder för att förbereda Sveriges jordbruk och livsmedelsindustri för EG' ['Measures to prepare Swedish agriculture and food industry for the EU'], Government Committee Report, Regeringskansliet, Stockholm

SOU (1994:119) 'Livsmedelspolitik för konsumenterna – Reformen som kom av sig' ['Food policy for consumers – The reform that did not materialize'], Government Committee Report, Regeringskansliet, Stockholm

SOU (1995:88) 'Den brukade mångfalden' ['The managed diversity'], Government Committee Report, Regeringskansliet, Stockholm

SOU (1995:117) 'Jordbruk och konkurrens – Jordbrukets ställning i svensk och europeisk konkurrensrätt' ['Agriculture and competition policy'], Government Committee Report, Regeringskansliet, Stockholm

SOU (1995:139–140) 'Omställning av energisystemet' ['Restructuring the energy system'], Government Committee Report, Regeringskansliet, Stockholm

SOU (1996:62) 'EU, konsumenterna och maten – Förväntningar och verklighet' ['EU, consumers and food – Expectations and reality'], Government Committee Report, Regeringskansliet, Stockholm

SOU (1996:65) 'Administrationen av EU:s jordbrukspolitik i Sverige' ['Administration of the CAP in Sweden'], Government Committee Report, Regeringskansliet, Stockholm

SOU (1996:184) 'Bättre klimat, miljö och hälsa med alternativa drivmedel' ['Improved climate, environment and health with alternative fuels'], Government Committee Report, Regeringskansliet, Stockholm

SOU (1997:25) 'Svensk mat – På EU fat' ['Swedish food – On an EU tray'], Government Committee Report, Regeringskansliet, Stockholm

SOU (1997:102) 'Mat och miljö: Svensk strategi för EG:s jordbrukspolitik i framtiden' ['Food and environment: A Swedish strategy for future EC agricultural policy'], Government Committee Report, Regeringskansliet, Stockholm

SOU (1997:150) 'EU:s jordbrukspolitik och östutvidgning' ['CAP and East European enlargement'], Government Committee Report, Regeringskansliet, Stockholm

SOU (1997:167) 'En ny livsmedelspolitik för Sverige' ['A new food policy for Sweden'], Government Committee Report, Regeringskansliet, Stockholm

SOU (1998:78) 'Jordbruk och miljönytta' ['Agriculture and environmental benefits'], Government Committee Report, Regeringskansliet, Stockholm

SOU (1998:108) 'Analysera mera' ['Analyse more!'], Government Committee Report, Regeringskansliet, Stockholm

SOU (1998:147) 'Effektivare hantering av EU:s direktstöd till jordbruket' ['More efficient administration of EU support to agriculture'], Government Committee Report, Regeringskansliet, Stockholm

SOU (2000:23) 'Förslag till svensk klimatstrategi' ['Proposal for a Swedish climate strategy'], Government Committee Report, Regeringskansliet, Stockholm

SOU (2001:68) 'Den nya produktionen – Det nya uppdraget' ['The new production – The new task'], Government Committee Report, Regeringskansliet, Stockholm

SOU (2001:77) 'Handel med elcertifikat – Ett nytt sätt att främja el från förnybara energikällor' ['Trade with electricity certificates – A new way of promoting renewable electricity'], Government Committee Report, Regeringskansliet, Stockholm

SOU (2002:75) 'Utrota svälten: Livsmedelssäkerhet, ett nationellt och globalt ansvar' ['Eradicate poverty: Food security, a national and global responsibility'], Government Committee Report, Regeringskansliet, Stockholm

SOU (2003:105) 'Levande kulturlandskap – En halvtidsutvärdering av Miljö och landsbygdsprogrammet' ['A living countryside – Mid-evaluation of ERDP'], Government Committee Report, Regeringskansliet, Stockholm

SOU (2004:4) 'Förnybara fordonsbränslen' ['Renewable vehicle fuels'], Government Committee Report, Regeringskansliet, Stockholm

SOU (2004:62) 'Handla för bättre klimat – Handel med utsläppsrätter 2005–2007 m.m.' ['Trade for better climate – Trade with emissions permits 2005–2007 etc'], Government Committee Report, Regeringskansliet, Stockholm

SOU (2004:133) 'Introduktion av förnybara fordonsbränslen' ['Introduction of renewable vehicle fuels'], Government Committee Report, Regeringskansliet, Stockholm

Svebio (2006) 'Svebios historia', Svenska bioenergiföreningen, Stockholm, accessed 16 March 2006 from www.svebio.se/Svebiohistoria.htm

Vail, D., Hasund, K.-P. and Drake, L. (1994) *The Greening of Agricultural Policy in Industrial Societies: Swedish Reforms in Comparative Perspective*, Cornell University Press, Ithaca, US

Virgin, I. (1998) 'LNB debattartikel 980701 Barsebäckskarusellen snurrar vidare' ['The Barsebäck carousel keeps on going around'], *Liberala Nyhetsbyrån*, Stockholm

7
Discussion: What Enabled EPI in Practice?

Måns Nilsson, Katarina Eckerberg and Göran Finnveden

A range of EPI measures have developed in Sweden over the last 10–15 years. The sector responsibility principle, established in the late 1980s, has become the principal backbone of EPI. Its implementation was then anchored in the late 1990s through the establishment of the national environmental quality objectives (NEQOs). Recent enhancements of environmental management systems in Swedish public administration have since strived to provide an effective system for implementation. Hence systems for EPI are in place, usually giving Sweden a high score in EPI benchmarking exercises (Jacob and Volkery, 2004). However, as was argued in Chapter 2, what actually goes on in the sectors does not by default mirror these institutional developments and policy aspirations. Therefore, we dug deeper. But what is our verdict? Has Sweden experienced EPI in practice? Have sector frames – and policymaking – changed? And if so, under what conditions?

This chapter discusses the major findings and lessons learned from our analyses in terms of promoting EPI as 'a first-order operational principle to implement and institutionalize the idea of sustainable development' (Lenschow, 2002, p6) As introduced in Chapters 1 and 2, unpacking EPI includes analysing what the environmental objectives that should be integrated are, when in the policy process this should be done, and how – with what methods? Moreover, the sector context for EPI is important, since how the sector is defined shapes the division of responsibility and setting of objectives. For instance, the 'sectoral regulatory capacity' (Hey, 2002) and the relationships between sectoral policymakers and interest groups is relevant. Our discussion is organized as follows: first, what are the current EPI outcomes, in other words does Sweden really have 'good' EPI or not? What environmental issues appear in the different analyses of EPI? Second, to what major factors can we attribute the EPI outcomes? What is the role of institutional arrangements for EPI? After these discussions, we provide a more generic reflection on theory and methodology, and on the concept of EPI as a process of learning. Chapter 7 addresses primarily an academic audience, while the more policy-oriented lessons are picked up in Chapter 8.

What are the Current EPI Outcomes?

This book has investigated EPI in practice in two sectors, energy and agriculture, which are responsible for some of the most severe environmental problems, but which, at the same time, are at the forefront of more active integration efforts nationally and at the European level. In addition, it has looked at bioenergy as a cross-cutting issue that at least partly provides a potential solution to some of those environmental problems. This part of the study has combined different methods: a 'sector environmental analysis' to assess both the direct and the indirect impacts of the activities in the sector; an analysis of the 'policy frames' that have emerged, how they have shaped major policy rounds and to what extent EPI as learning has happened; and a content analysis on how different environmental issues and concerns have shown up in this process. Hence it has addressed both EPI in the process of policymaking and actual environmental outcomes, but has not been concerned with establishing causality between the two.

It was shown in Chapter 5 that, over time, frames have evolved and sometimes also shifted significantly, thus accommodating new understandings of how the sectors work and what they, along with their designated policy landscapes, are expected to achieve. These new framings, and the way in which they have influenced policy, indicate that conceptual policy learning has taken place. Furthermore, partial EPI has occurred in both sectors as certain concerns about environmental sustainability have become part of the dominant sector framings. However, these processes have played out differently in the two sectors.

The energy sector has adapted its policies and strategies from a national planning frame (energy-as-infrastructure) to a market frame (energy-as-market) with wider international boundaries. At the same time, the climate change agenda has become strongly engraved in the sector mainstream and all main actors accept it as a key priority of the sector, which is a significant sign of EPI progress over the last 15 years. In this field, policy instruments abound: carbon taxes are combined with tradable emissions permits and tradable certificates. Indeed, these policies, introduced since the early 1990s, have led to a conversion from fossil to renewable fuels (and to waste incineration) in thermal heat and electricity production. However, environmental concerns in energy policy in Sweden today are largely limited to climate change and associated non-renewable resource issues. It appears as if the attention to climate change has crowded out other issues of relevance, such as biological diversity, landscapes and air pollution. Air pollution is controlled through the environmental quality norms in the Environmental Code but has, for the last ten years, not been an active policy concern in energy policy.

Despite climate concerns being prominently integrated, the dominant framing of the energy sector over the last couple of years has been energy-as-market. The deregulation of the energy markets has been driven by a political process in parallel with, but more or less disconnected from, the political process surrounding the green transition agenda. Because a market framing suggests that governmental intervention should be minimal, this has constrained the opportunities for the state to intervene with policies to correct market failures. The smaller government agenda has taken over in the policy discussions concerned with energy markets, despite widespread principled agreement within the market frame of the importance of controlling for externalities of economic activities.

Agricultural policy has been ruled by the Common Agricultural Policy (CAP) since the EU membership in 1995. According to the CAP reform in 2004, direct production support is being transferred to support improved land management, including the environment, animal health and rural development. EPI has materialized particularly within the 2000–2006 Environmental and Rural Development Plan (ERDP) for Sweden, which will be replaced with a new Strategy for Sustainable Rural Development for 2007–2013. The support mechanisms set agriculture somewhat apart from more deregulated sectors. In addition to traditional instruments such as regulations and taxes, significant funds are dispensed as 'carrots' for environmentally friendly production.

Two different agricultural policy tracks emerge. One relies on deregulation and open markets, embracing the ideas of agriculture-as-entrepreneurial-practice and agriculture-for-security. This has been the main track of the CAP and promoted by established agricultural interest groups. The second policy track is pursued through the ERDPs, and this is where agriculture-for-sustainability is strong. Here several environmental concerns have for some time been central. Heavy significance has been put on landscape issues, in terms of both cultural and natural values, supported through a series of government programmes that were already in place in the late 1980s (and thus before the influence of the CAP). Policies are also in place for polluting emissions to air, ground and water, although there are continued problems with environmental impacts. The farmers' interest groups have become increasingly engaged in promoting environmental management practices within this field. However, the agricultural sector's potential to contribute to climate change mitigation has been largely absent.

If we compare the general manifestation of EPI in the two sectors it appears that many environmental issues have become embraced in the agricultural sector, whereas in the energy sector only the climate issue has become integrated. On the other hand, climate issues are now firmly included in the energy sector 'mainstream' framing, whereas the environmental issues in the agricultural sector are contained in the ERDP. Such containment also occurred for the energy sector during the deregulation process in the 1990s, which was kept separated from the green transition agenda that unfolded during the same time period. However, as some side effects of deregulation have surfaced (SOU, 2005:4), the energy-deregulation debate and energy-as-market framing have become strongly exposed to both environmental and other public-good concerns such as supply safety.

Bioenergy addresses the problem of how to reduce the use of non-renewable resources in both the energy and agricultural sectors. However, although political signals have been manifest at both national and EU levels, a lack of coordination between the two sectors, and a slow response from the agricultural sector, have constrained the development of bioenergy resources. Recently, as a result of changes in the EU's ambitions regarding renewable energy and the CAP reform, new Swedish coordinative measures have been undertaken. A new bioenergy trend is emerging at the national as well as at the European level, once again framed as a security issue but also emphasizing sustainability issues.

So far, we have discussed EPI as if 'the environment' were a concept of low complexity. However, different environmental problems might well clash and be difficult to resolve at the same time, for example an open landscape with cattle grazing and reduced loading of nutrients to water systems in agricultural policy, or protection of pristine rivers and extended use of fossil-free hydropower in energy policy. Thus solving one environmental problem often leads to unravelling new environmental problems and risks, making the EPI quest a classical 'wicked' problem (Rittel and Webber, 1973). For instance, bioenergy can be seen as a solution to environmental hotspots in the energy and agricultural sectors by reducing the environmental impact of energy use in both sectors (provided that clean technologies are used), as well as increasing biodiversity. However, bioenergy production in agriculture might actually also *increase* certain impacts such as nutrient leakage and a less open landscape. Hence there is also a need for coordinated evaluations of bioenergy policy in order to attain a holistic picture enabling a correct judgement of the environmental net result from increased bioenergy use.

As discussed in Chapter 2, one must therefore unpack what is really meant by 'the environment' when analysing the implementation of EPI. We will turn to this discussion now. Chapter 4 assessed the environmental impacts of the agriculture and energy sectors. From previous studies we know that these sectors are among the most important, from an environmental perspective (Huppes et al, 2006; Palm et al, 2006). The sector environmental analysis identified important environmental problems for the two sectors. In the energy sector, the use of non-renewable resources, climate change and air quality aspects, and the toxicity of fossil fuels showed up as potentially the most severe environmental impacts. The problems largely originate from the use of fossil fuels. Solutions could potentially be found by improving combustion techniques, but also through fuel substitution and reduced energy consumption. One of the most feasible options is the increased use of biofuels. In the agricultural sector, eutrophication, use of non-renewable resources, climate change, biodiversity and toxicity were identified as potential 'hotspots'. Major contributors to these problems include activities that occur within the sector (for example farming practices, emissions from cattle and nutrient leakage from soils), as well as those that occur within the energy sector but are the result of decisions made within the agricultural sector (for example the use of non-renewable resources as fuels for transport, heating and electricity). Hence the importing of food and the behaviour of consumers seem to be as significant to the discussion of environmental impacts as farming techniques. The sector analysis demonstrates that the current understanding of what a sector entails limits the awareness of the full environmental effects related to the sector. A broader life-cycle perspective on sector activities shows that many of the major impacts occur outside what is normally perceived as the sector. This has most likely limited the effectiveness of policy and points to an allocation problem that restrains the EPI potential. That environmental issues are indeed considered to be serious and problematic seems to be a necessary but insufficient condition for integrating them onto the sectoral policy agenda (Engström et al, 2006). Through the content analysis presented in Chapter 5 we could also observe the path dependency and limited attention capacity of the policy system. It appears that the agenda can only hold a limited number of issues and that in most cases once issues have made it onto the agenda, they remain there, regardless of whether other ones emerge as more important.

The mixed patterns of EPI lend themselves to a more in-depth discussion of potential causalities and relationships. In the following we distil some discussion points that give a better understanding of what factors have been decisive in shaping the observed EPI patterns.

What Key Factors Affect EPI Performance?

Chapter 5 showed that at certain times, actors adjust their framing in conceptual learning processes. At the aggregate policy level, this may ultimately manifest itself as changes in policy, including assimilation of new environmental policy objectives such as climate change. What types of underlying factors facilitate such policy learning and subsequent policy change? The study presented in this book confirms what literature on policy change and learning has been saying for many years: external driving forces are necessary triggers, and perturbations such as a crisis or an exceptional event can provide a window of opportunity for frame change (Sabatier, 1998). 'Policy windows' emerge in certain moments in time where the policy agenda is open for systemic change (Kingdon, 1995; Lebow, 1984). For instance, both the Chernobyl and Three Mile Island nuclear incidents had profound implications on Swedish energy policy, and the development of the EU completely recast Swedish agricultural policy. This study suggests that policy windows are not always abruptly opened but can also evolve slowly, as happened in relation to the change in the energy market and its configuration of actors over a decade. It is also useful to discuss factors at different levels. In the next section we discuss the implications of some of the overarching drivers at the international level. In the section thereafter we look at the domestic political aspects and conditions.

The drivers: Market reform, the EU and global commitments

The evolutions of the international policy context and commitments vis-à-vis the EU and international conventions have been important drivers for EPI nationally. However, the mechanisms involved are complex and contradictory. On the one hand, the EU has been considered a positive driving force for environmental protection across Europe (Baker, 2000; Jordan, 2002). Sweden and other front-running member states have pushed for stricter European environmental policy, often quite successfully (EEA, 2005b; Skou Andersen and Liefferink, 1997). On the other hand, the prominence of the environmental agenda in the EU fluctuates heavily with the interests and powers of individual members of the Commission and the Councils, which also has repercussions for the Swedish agenda. And while the positions on environmental policy in the European environmental legislature might be as far reaching as those of the Swedish government, it appears that the integration of environmental policy into economic sectors has been far less advanced. Indeed, it has been argued that agricultural policies in Europe, over which the EU has competency, have had much more difficulty becoming green than, for instance, energy policy, where national governments have competency

(Nilsson and Nilsson, 2005; EEA, 2005b). With Swedish society as a whole being more tuned in to environmental issues than the European average, EPI in economic sectors faces fewer normative constraints. The implications of the EU membership are therefore contradictory: taking a broad view, EU membership did not positively contribute to EPI in Swedish agricultural policy, whereas it was a positive driving force for climate integration in energy policy. In the latter, due to binding commitments and forced transposition of European directives to implement emissions trading, there was little scope for challenge by vested interests. Through the process of preparing the transposition of the directive, sectoral actors developed acceptance of and responsibility for the climate issue (see also Nilsson, 2006). The evolution of Swedish bioenergy policy during the last 15 years was spurred by international events, rather than domestic initiatives, and it therefore fits the pattern of international influences. Furthermore, the most rapid developments in bioenergy policy have occurred not on environmental grounds, but on economic and security grounds.

International influences and policy transfer can also occur informally and through changing norms. The international market paradigm, which lies at the core of the EU and the 'internal market', has a complex relationship with EPI. On the one hand, the primacy of economic markets and the growing confidence in the role of market mechanisms in shaping effective policy in the international arena spilled over to the national policy debates. This, in combination with the economic crisis of the early 1990s, paved the way for a revival of economic growth pursuits, with the apparent result of social and environmental policy considerations being less prioritized. This was reflected in the policy bills of the time, which praised the need for economic growth and competitiveness. On the other hand, the new market orientation and internationalization was a challenge to the existing dominant frames and networks, and contributed to breaking up hegemonic frames in both the energy and agricultural sectors. As such, it might be considered a necessary first step for subsequent integration of environmental values. The internationalization of policy and European integration created opportunities for EPI by opening doors for new policy measures, closing doors on some old ones, and allowing new synergies to be explored between, for instance, supply security and environmental protection.

However, to associate the market-orientation trend with the EU entry is overly simplistic. Market framing of the sectors emerged in energy-as-market and agriculture-as-entrepreneurial-practice even before the EU entry was high on the agenda. Furthermore, the EU entry in 1995 actually took the Swedish sector policy in opposing directions in terms of market orientation: a 're-regulation' of the agricultural sector, which had just become deregulated in the early 1990s, and a deregulation of the energy sector.

EPI within Sweden might also have been facilitated by Sweden's entry to the EU in another way. Swedish actors discovered that their national interests were fairly similar to the broader international interests. Thus Swedish industrial and environmental interests have often formed joint positions about environmental protection requirements in arguments against industries and governments in countries with a weaker environmental protection tradition or with other legislative traditions:

We see a united Swedish front towards the EU – old combatants have joined hands as they have a nice external enemy to fight. The EU has a tendency to regulate in detail, whereas in Sweden we have more of trust, with concessions in retrospect and similar, that amounts to a less legalistic view and more of a consensus model. (Interview, industry representative, 2004)

The process of alignment between the environment and industry was further facilitated by the international evolution of the green political agenda, moving from anti-growth in the 1980s to ecological modernization in the 1990s (Hajer, 1995). In the late 1980s, the Green Party was committed to inducing a radical reinterpretation of our lifestyle and a new 'social contract' whereby happiness would no longer be sought through increasing economic activity, but rather from more leisure time and lower material aspirations. However, this no-growth agenda failed to find popular support beyond a very small group of voters, and as the party became part of the political establishment in the 1990s it was more or less dropped. Ecological modernization provided the possibility of combining environmental ideals with mainstream economic welfare aspirations, which was also attractive for Social Democrats. This was probably strategic behaviour as much as conceptual learning. Like the Green Party, they saw that in order to gain the public and interest-group support necessary to achieve potent environmental policies it was necessary to play along with the growth agenda rather than fight it.

Another important driver for EPI is the development of knowledge about the environmental problems themselves. One apparent example is climate change. During the period studied here, the evidence and consensus on the climate change issue has increased, as manifested in the reports of the Intergovernmental Panel on Climate Change (IPCC). The gradual accumulation and strengthening of the knowledge base has undoubtedly been an important driver for the integration of climate change issues in energy policy. In the agriculture sector, knowledge gained on biodiversity and eutrophication issues are other examples where communication of scientific results have informed policymakers and affected policies. However, the development of the knowledge base in itself is insufficient. Knowledge has increased or changed in other areas as well. One example is the revival of air quality as an important environmental issue. This was caused by the insight gained within the last decade that fine particles can be a severe health problem in Sweden (Forsberg et al, 2005). The fine particles are both emitted directly and formed as secondary products from other air pollutants, such as nitrogen and sulphur emissions. As shown in Chapter 5, these insights have not, however, yet had any major impact on policy documents in the two sectors.

The domestic context: Actors, networks and opinions

Successful environmental responses have been said to depend on whether actors that stand to bear the costs of the intervention, and are hence opposing it, are or can be organized into influential coalitions (Skjaerseth and Wettestad, 2002). Chapters 4 and 5 show the flip side of this argument, that successful environmental responses depend on whether there are strong actors within the sector that might benefit from such a response, by being part of the technical solution or by being negatively affected by current environmental problems (Engström et al, 2006). For example, supporting coalitions within the sector were conducive to climate change goals in the energy sector: a) the nuclear industry benefited from pushing climate onto the agenda and b) the power producers could make massive profits on their hydro and nuclear power investments due to the electricity price effect from the emissions trading scheme.

Actor coalitions can also be consciously shaped. One example is how the government actively created constituencies through decades of research and development (R&D) support for bioenergy programmes. Suddenly there is a large group who have built their careers or competencies around a technology and therefore have a strong interest in seeing it through. The same thing happened on a much larger scale for nuclear power in the nuclear development programmes in 1950s and 1960s. Thus the sectoral policy networks contain different subsystems that can be nested within each other. For instance, the nuclear energy subsystem is nested within a broader energy supply subsystem. It can also overlap or build coalitions with other subsystems, such as the air pollution subsystem.

The importance of having actors within the sector that can benefit from changes also implies that the market structure of the sector can be important for the possibilities of EPI. If the market is close to monopolistic, or dominated by a few similar actors, it may be more difficult to find actors who can benefit from a change and thus less likely that an issue is placed on the policy agenda. The cohesion of sector networks determines the extent to which outside actors, ideas and solutions can enter (Daugbjerg, 1998). Although Swedish political administration has traditionally been based on a corporatist model, larger changes in network configuration occurred, in particular in the energy case. As the energy policy network changed in the late 1990s, the traditionally dominant industrial actors could not restrict access to new actors, which broke up the power balance and enabled the entry of new ideas and knowledge. However, whereas the energy sector has a more open network with shifting coalitions, the agricultural sector has retained a more hegemonic corporatist structure, although it too has opened its doors somewhat to new actors. The different network characteristics have probably influenced the differential EPI in the two sectors: the more closed agricultural sector has experienced contained 'within-frame EPI', whereas the energy sector has experienced EPI through a process of bridging different frames (Nilsson, 2005).

The political balance in parliament, as a result of elections, exists in parallel with the power structure between actor coalitions in the policy network. As the governing party relies on collaborating parties for a parliamentary majority, the relative weight of different priorities might be different (Weale et al, 2000). Most observers agree that the overall political collaboration between the government and the Greens has been a factor for pushing environmental concerns in all sectors: 'The Ministry for the Environment is a weak actor. This is compensated for by the political situation today' (Interview, SEPA official, 2003). However, here we also have different implications: in energy policy, political negotiations at the party-leader level have been prominent in all major policy decisions, whereas agricultural sector policy preparations are far less politicized and not subject to the same types of dedicated party-political negotiations (aside from the annual budget negotiations). In agricultural policy, political parties have held considerably lower profiles.

Political leadership has been important for EPI in Sweden. The launching of the 'green people's home' in 1996–1997 (see Chapter 1) paved the way for turning the major policy programmes into green programmes: the environment and rural development programme in agriculture; the transitions programme in energy. This connects to a common 'trash-can' variable in political science: political will that people tend to throw in when they find no other explanations for a certain political outcome. Political will is a complex phenomenon. The signals from the top are necessary as a motivator, but government is fragmented, and different ministers have different political interests; different authorities are heavily influenced by who is in charge and who is assigned as their counterpart (supervisors) in the ministry. At the same time, political will in the sectors can also be considered a result of, rather than condition for, EPI. A reframing whereby environmental concerns become a natural part of the agenda is part of the interest formation that shapes political will.

What about public opinion and EPI? The Swedish public has a strong level of environmental awareness, and environmental concerns rank higher among the Swedish public than in most countries (Inglehart, 1997). The public saliency of the environmental agenda provided the leverage for green ideas to reach the political mainstream in the late 1980s in Sweden (Eckerberg, 2000; Lundqvist, 2004) and elsewhere (Lafferty and Meadowcroft, 2000). However, in the 1990s, media attention and, in consequence, public interest faded (Holmberg and Weibull, 2001). This created problems for pushing environmental agendas in sectors since politicians are very mindful of public opinion:

> *And here the Greens tried to push the agenda, but others have felt that if we go too far here we will get public opinion against us. [...] When lobby groups take the voters away from the parliamentarians, then they have difficulties standing up for their cause. So even if they believe in the cause they will say, 'No we do not buy what you are saying'.* (Interview, leading Social Democrat politician, 2004)

Today environmentalists, industry and government tend to follow the same rhetoric. Corporations promise to be good environmental citizens and work with NGOs. NGOs accept industry's principal positions, for example regarding fair competition. That different actors now follow a common environmental rhetoric may, however, be an illusion. In controversial issues, environmental arguments may be used both for and against a contested project, and it may seem as if environmental arguments serve only as rhetorical tools in the argument (Isaksson, 2001). However, these environmental disputes are not just matters of empty rhetoric. The different actors may, for example, be talking about different environmental aspects, giving priority to some over others. They may also refer to the distribution of the environmental 'goods' and 'bads'. Thus what may appear as environmental rhetoric may involve real political conflicts (ibid). The possibility of using environmental arguments both for and against the same project also points to the need to use assessments that can be used as a common starting point. However, it also points to the difficulties in developing common assessments to come up with direct 'answers' where both sides want to use assessments as 'strategic ammunition'.

What is the Role of Institutions for EPI?

As discussed in Chapters 2 and 6, the institutional and organizational landscape is often referred to as key to how environmental issues are discussed in preparations for policy change. We have investigated how EPI practice relates to institutional arrangements in Sweden. In the following section we separate the general features of the institutional arrangements surrounding policymaking from the specific institutional measures introduced to promote EPI. The discussion below shows that the potentials for and barriers to EPI vary substantially at different levels or stages, such as the framing of the problem, the agenda setting, follow-up and assessment processes, or when negotiating trade-offs at the decision stage.

Corporatism, sector definitions and EPI

There are reasons to expect that Sweden would have better conditions for integrated policymaking than most other countries. Sweden's policy style has been characterized as one of rationality, clarity, corporatism and compromise (Einhorn and Logue, 2003). Openness, size and political institutions – including the long period in office of the Social Democratic government – are said to explain this political style. Its institutional organization enables the government to develop fact-based policy proposals via a process of consultation and compromise and to implement them efficiently. Various interests, including environmental groups, are consulted through this mechanism. Moreover, the rule of public access to official documents facilitates exchange of information from government bodies to those affected. In general, therefore, the channels between government ministries and agencies and various interest groups are well established in Swedish policymaking, which should facilitate EPI. However, earlier studies have shown that, in practice, environmental interests are not always present in all phases of policy formulation (Uhrwing, 2001).

The corporatist institutional culture has an ambiguous relationship with EPI. Earlier empirical work has suggested that corporatism promotes environmental performance, notably through this institution's ability to provide public goods (Jahn, 1998; Jänicke and Weidner, 1997). However, environmental performance in general is not the same as our definition of EPI (how sectors reframe). When it comes to EPI as reframing of sectoral processes, our studies show a more complex relationship (see also Scruggs, 1999). On the one hand, the corporatist approach can be an impediment to EPI since it reinforces the network structures and dominant patterns and frames. In both our cases, EPI was initially triggered by the breaking up of corporatist networks. On the other hand, when environmental interests are part of the 'corporation', the consensual culture associated with corporatism has been largely conducive for EPI, which many interviews testify: 'Here environmental organizations are part of the discussions of consensus. If you are let in, you get a responsibility and you act differently. If you are not, all that remains is to scream and shout' (Interview, SEPA official, 2003). Still, the corporatism might be problematic for conceptual learning in a wider sense because of a tendency to conformity:

The discussion at the time made me extremely depressed. Because it does not matter if you are right! Sweden is a small conformist country, and there is something in the economic-political discussion that is politically correct. If you do not have the politically correct view, you have no influence because no one cares what you say. That was my experience. (Social Democrat politician Anders Ferm quoted in DsFi, 1999:27, p66)

On balance, the effect on EPI from corporatism is that a relatively narrow environmental agenda can be pursued, although environmental problems, such as those discussed in the NEQO framework, are highly multidimensional. For instance, we have seen that in the agriculture sector, farmers' interest organizations have increasingly embraced environmental values and created their own agendas for some of them (Chapters 5 and 6). The consultative and compromising policy style has its merits, but also tends to limit the potential influence of those interests that are not involved.

Corporatism is also related to the way that the sector is defined: who is part of the sector and who is not? This affects the potential for EPI in several ways. First, we saw from the sector environmental analysis that a number of environmental problems, caused by activities from within the sector, did not appear clearly on the agenda for discussion in the various policy rounds. This was particularly true for energy issues relating to transport, and to the use of non-renewable resources in the agriculture sector – problems were perceived as lying outside of the sector as it is traditionally defined. The conceptual understanding of the responsibilities within the sector was, therefore, that the problems did not really need to be solved from within the sector itself. A similar argument could be raised for the role of consumers, who can 'vote with their feet' and choose ecological products or not, thereby having influence through the market mechanisms. In 2006 the Minister of Agriculture declared that Sweden should aim for 20 per cent ecological farming through consumer pressure (by educating consumers through measures by the state). Although this statement highlights the growing importance of consumers, they have rarely been included in the agricultural sector's own definition of their sector. What then determines a 'sector'? The conceptual answer lies in collective action: when actors with similar interests strive to get their voices heard, and form networks to create the capacity for change, eventually – if there is sufficient common interest and public pressure – this also becomes institutionalized in government organization. In practice, however, the 'sector' is defined and generally accepted when it is given a special allocation in the government's budget. This indicates that strong framing has occurred in how the sector is perceived, and it is difficult to introduce measures that cut across the sectors since none of the established sector agencies is clearly accountable.

In our analysis, we found policies that do not fall easily into a single sector and thus become difficult to handle. The most striking example is bioenergy, cutting across agriculture, energy and forestry sectors. Bioenergy could expand when sectors coincided in their setting of policy goals and devising of measures to support bioenergy production. But when the CAP changed the rules of support to bioenergy production, its rapid development came to a drastic halt. The policy goals for the two sectors were no longer coordinated. Bioenergy received rhetorical attention in both sectors but was promoted in practice only by the energy sector. Another example is policy for rural development, which struggles in a no man's land between regional and economic development/industrial policy and agriculture and environmental policy. The restructuring of the agricultural agencies in the mid 1990s was partly justified by a need to better integrate policies for rural development, and the national ERDPs have also aimed for sustainable rural development. However, there is little integration across the units for agriculture, environment and regional development in the everyday handling of these matters in the regional governmental administrations. Partly this can be attributed to different professional cultures among those units, and perhaps also to the heavy workload in the agricultural units implementing and controlling the CAP measures, which has rendered it difficult to engage in outreach and coordination activities.

The competencies of ministries generally reflect the perceived needs of the sector, and the building up of new skills to reflect EPI moves slowly. We have seen that public officers in charge of environmental issues are often working in relative isolation from the mainstream, even if there is increasing lip service paid to environmental goals. There is also a tendency for only those environmental issues which suit the interests of the particular ministry or agency to be brought onto the agenda, while others get lost. The support for ecological farming is a case in point: the MoA has increased its emphasis on this as a way to maintain open landscapes and reduce toxicity; however, the continued problems of eutrophication will not automatically be solved through such measures and are therefore downplayed in the discussions.

The committees of inquiry

Committees of inquiry (introduced in Chapter 1) are influential in the construction of policy and are typically widely cited in government bills. They are the most important 'bundling arena' between science and policy (Jasanoff, 1990) and therefore should also constitute the key mechanism for integrating environmental knowledge into policymaking. Is this mirrored in reality? In some cases, committees seem to function relatively well for EPI purposes. We have seen that, through committee processes, mainstream economic interests and actors have come to accept relatively powerful environmental policy measures and objectives, for example in the committee that prepared the NEQOs or the implementation of the EU emissions trading directive. If managed properly, committees have a trust-building effect and, lower-profile committees in particular, also provide a relatively depoliticized forum for policymakers to take stock and learn. However, although committees have considerable potential, there are problems in practice that limit their effectiveness in supporting EPI.

First, the committee is strongly controlled by the ministry that is the 'client' of the process. The ministry gives careful instructions regarding the scope and content of the committee investigation, who will participate, and what impact assessment will be done. EPI within committees is restricted by the competencies of their members. This creates a strong framing effect that prevents the committees from engaging in conceptual learning processes in terms of finding new perspectives on, agendas for and solutions to the policy issues at hand. Although there are generic instructions in the form of a 'Committee Guideline' (Ds, 2000:1) that highlights cross-cutting issues, the specific instructions are the ones that really matter. Furthermore, and surprisingly, the environment has so far *not* been among the important cross-cutting issues described in the guideline. Environmental interests seldom come in at the early stages of discussion, but are consulted only when important decisions have already been made, thus restricting the policy alternatives to be assessed. EPI tends to lose out when major actors have positioned themselves and therefore are not open to new information.

Second, committees have varying set-ups and mandates. Important and strategic policies are usually preceded by larger 'parliamentary' committees, often called 'Commissions'. These include representatives from all political parties, which should allow the development of viable political solutions, in other words finding a majority agreement that will hold in parliament. Such committees tend to become highly politicized, however, which can have a detrimental effect on learning and the potential for EPI. The agreements they make are frequently overthrown in subsequent bill preparations, as was the case in the Energy Commission 1995 and the Climate Commission 2000. Therefore the learning and EPI that might take place in the committee process is sometimes discontinued and not institutionalized at higher levels.

Third, it can also be questioned to what extent the smaller expert-driven committees really constitute arenas for policy learning as long as the experts are not taken both from within the sector and from external actors, such as environmental and other social partners. Learning is promoted by cross-fertilization of ideas and acquiring new insights, which is less likely to occur if already-established frames dominate the committee work. Thus there is a balance to strike between searching for those experts who are trusted by the political decision makers and those who may contribute to EPI.

Ministries and political decision styles

Why does the learning sometimes get lost at higher levels? Let us have a closer look at the ministry and political levels. As explained in Chapter 1, the Swedish central government is, simply put, separated into the parliamentary system that formally takes the political decisions, the Government Offices (ministries) that prepare policies for approval in parliament and the Central Governmental Authorities (agencies) that implement policy. Questions of policy and strategy should, according to the Swedish administrative system, be handled within the ministries, whereas agencies should implement and assist. However, in some senses the system seems to work backwards. If one considers strategic thinking, learning and a longer-term outlook as important elements of policy formation, in many ways agencies are better equipped. The more interesting, unbounded thinking is usually presented in agency reports. Why is this the case? Officials tend to talk about capacity constraints; ministries are generally caught up in day-to-day dealings and Brussels affairs and have limited time to reflect or think ahead. Indeed, since the EU entry, coordination of policy as an important part of EPI has become increasingly difficult with raised expectations for quick reactions to complex problem areas. A massive increase in ministry staff resources over recent years suggests that serious capacity constraints have been addressed, but some observers point to incentive structures and career paths in the government as a key constraint in the government's functioning. Desk officers change jobs relatively frequently and tend to be rewarded more for taking new initiatives and developing new bills than for gathering evaluative knowledge or ensuring that implementation and coordination is effective (Molander et al, 2002; Nilsson, 2006).

In addition, we would like to draw attention to the strategic-political nature of ministry-level policy processes. Whereas agencies are able to collaborate between, for instance, sectoral agencies and environmental agencies and to work on joint reports and initiatives, ministries have a more bounded perspective: they are expected to watch over their specific interests, or mandate, and let other ministries deal with their own issues. Thus, in very simplified terms, the MoF is only concerned with tax revenue and budget balance, the MoI is only concerned with the competitiveness of Swedish industry, and the MoE is only concerned with environmental protection at all cost. Therefore their inter-ministerial preparations for policy coordination, which are by law required – the 'joint drafting' procedure (in principle a very powerful mechanism for EPI) – becomes more of a strategic game and positional war than collaborative problem solving, with little scope for learning or reframing as EPI requires. In particular, bioenergy policy, as a multi-sectoral case, demonstrates that these coordination functions need to work better. Looking at Swedish government from a policy coordination perspective, 'negative' coordination is the major feature of the traditional system for policy preparation, through the use of 'joint drafting'. As soon as one ministry's policy initiative may have an implication for another ministry, it needs to get involved in consultations. However, as we have seen, this has not been enough for policy integration, but is more about avoiding stepping on each others' toes. The achievement of sustainable outcomes across sectors depends on 'positive' policy coordination where all sectors strive for the same objectives.

Finally, governmental policy overall is, to a large extent, determined in the annual budgetary process and associated negotiations. These negotiations tend to become very short-sighted and unpredictable processes where the interest in knowledge and lessons from experience, environmental or other, is difficult to maintain. Whether environmental issues get a strong outcome, then, simply depends on the relative power and stamina of parties and persons with strong environmental values (for example the Greens, as discussed in the previous section) and also to what extent other actors (for example the Prime Minister himself) share environmental values. While there is evidence of such environmentally friendly outcomes, it is clearly a very unstable system and it cannot really be described as EPI in terms of sector reframing. This does not, however, preclude EPI in terms of learning in later stages. Once decisions have been taken and implemented, learning may occur as the impacts of the decision become known.

Specific institutional measures for EPI

The 'institutionalization of environmental policy', in other words the extent to which formal institutions be put in place for EPI – how does that impact on EPI? Recently, the traditional model of policymaking has been complemented with initiatives to enhance the potential for coordination and integration. As detailed in Chapter 1, the 1988 bill 'Environmental policy for the 1990s' served as a starting point by establishing the principles of sector responsibility. In the mid to late 1990s, it took off in operational terms, and the creation and gradual accumulation of environmental integration measures has since created a relatively powerful portfolio including the NEQOs, assessment procedures, environmental management systems and environmental coordination bodies. Horizontal initiatives are thus combined with hierarchical ones such as management by objectives and sector accountability schemes. Tensions may arise between horizontal coordination, whereby actors interact and develop a common agenda in a bottom–up approach, and strong vertical linkages, whereby the political leadership exercise a hierarchical control, but in Sweden the government tries to do both at the same time. In the following section, we will discuss each of these measures in more detail.

Quantified environmental objectives and sector responsibility

In the late 1990s the parliament adopted a new target-oriented environmental policy approach including the 15 (now 16) NEQOs. These integrated the existing 170 environmental targets into a new scheme (Prop, 1997/98:145). In 2005 the system was reviewed and further enhanced in the bill 'Swedish environmental objectives – A joint mission' (Prop, 2004/05:150). The set-up of this NEQO system and associated management systems have been positive for EPI. Through its management system, and supported by the sector responsibility principle, NEQOs carry the potential to interact with other policy processes. In a sense the NEQOs have made sector responsibility operational, and the assessments of distance to target have a simple appeal (Figure 7.1). Three aspects in the development of NEQOs appear particularly important for this positive spin. First, through the substantive targets and systematic follow-up, accountability and professionalism are promoted. Second, through the process of developing and anchoring the NEQOs in the mainstream sectors, and through designating the agency heads as members of their governing council, political will and ownership is promoted. Third, the major political agreement in parliament on the core system (all parties except the Conservatives endorsed the regime) induced a real momentum in the process.

Therefore the European Environment Agency's (EEA's) verdict of the NEQO system in its most recent report on EPI seems overly pessimistic:

> *Sweden's experience with management by objectives demonstrates some of the limitations of NPM (new public management), suggesting that it is not simply an opportunity but also a potential threat to EPI. The more rigorous the strategic management initiative, the more difficult it may be for environmental or other horizontal objectives to be 'added on' subsequently.* (EEA, 2005a, p7)

The EEA holds that problems emerge from the lack of political prioritization of NEQOs, the lack of budgetary mechanisms associated with them and the lack of power on behalf of the Swedish Environmental Protection Agency (SEPA) to intervene in the sectors. These problems do exist, but on balance our study suggests that the NEQO system has been a useful step forward for EPI, in terms of ownership, accountability and organizational capacity for dealing with environmental issues. In particular, the NEQOs have helped to raise awareness about the

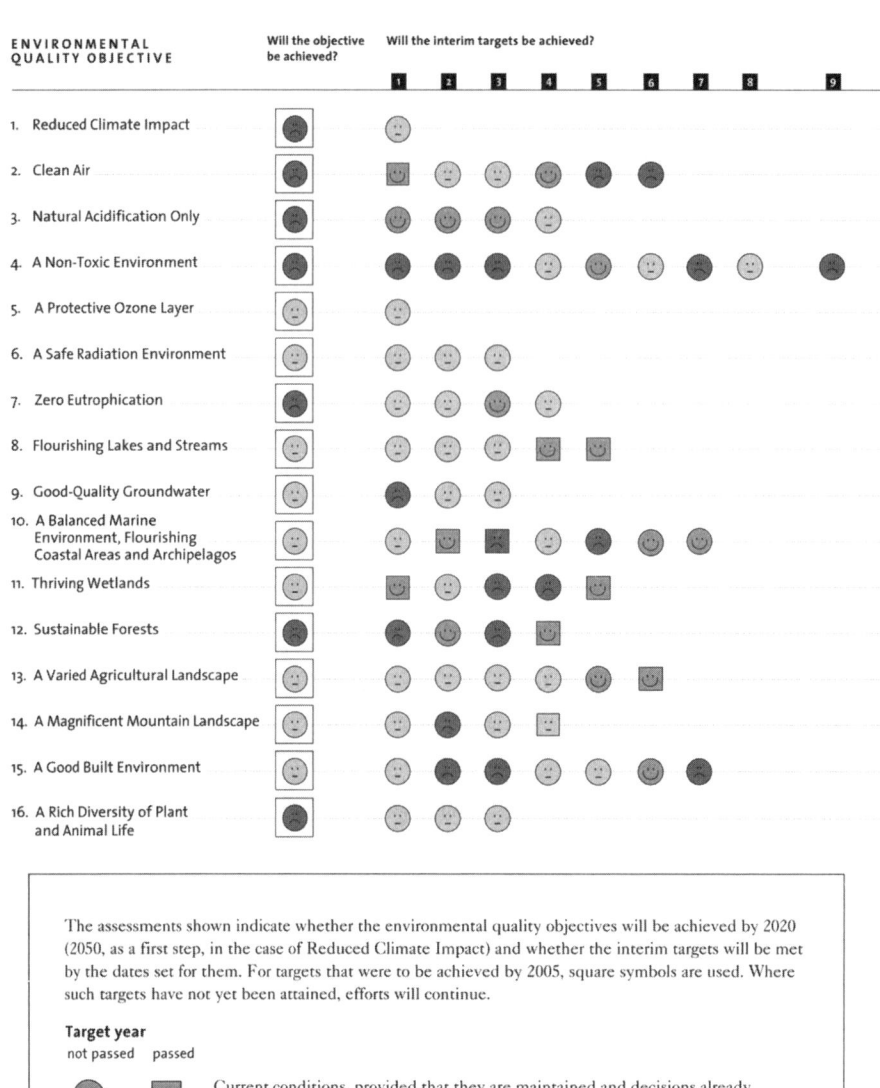

Source: www.miljomal.nu

Figure 7.1 *Simple assessment of whether environmental objectives and interim targets will be reached*

need for EPI in the various policy sectors, although there is still a clear tendency to regard the NEQOs as environmental policy rather than the respective sector's policy goals. Furthermore, even if the NEQOs may not be regarded as very important in short-term decision making, they are often considered more in rhetoric about future policies.

Assessment systems

From an environmental point of view, procedures for impact assessments are generally weak and poorly implemented today. In both energy and agricultural policymaking, assessments of policy alternatives have been made primarily at the initiative of the committees of inquiry. Since committees are normally governed from within the sector, economic assessments have been prioritized and environmental effects have been investigated only marginally. Environmental issues are not even mentioned as one of the important cross-cutting issues in the generic instructions to committees. The narrow framing of the committee assessment work also means that assessment of new practices proves particularly difficult, as illustrated both by the GMO question in agriculture and the cross-cutting bioenergy issue. When there is not yet established knowledge to be drawn from, and the potential risks are cumbersome to measure, there is less room for learning.

As part of the NEQO and EMS schemes, environmental assessment procedures are currently also stipulated in the Government Offices (the ministry level). However, this procedure has been limited to a one-page checklist where the desk officer in charge marks whether the proposal has negative or positive impacts on each of the 16 NEQOs. Furthermore, these assessments are not available for public scrutiny but are considered internal working documents. A recent evaluation showed that officials in charge of the impact assessment spent, at most, one half day on the assessment, and in most cases no more than one hour (Miljödepartementet, 2003).

The weakness of the assessment procedure has meant that important decisions have not systematically considered environmental issues, and environmental accountability has become dependent on political will. What is needed is a very strict quality control and upgrade of the assessment procedure. Recent communications from government have suggested that better assessment procedures are under way (Skr, 2005/06:126). Stricter guidelines for assessments, including minimum requirements, will be adopted, a change following and inspired by the similar movement in the EC. To what extent these changes will be enough is still not possible to judge.

Environmental management systems (EMSs) for the governmental administration

EMS schemes have been implemented in the governmental administration since 1997. At first, they focused on agencies' internal office work, rather than relating to the substance of policy decisions. Agencies and ministries managed their travel better and saved paper and energy, but what about the policies they decided upon? In recent years, there has been a systematic attempt to move towards more effective systems that relate more to the substance of decisions, not just office work (Miljödepartementet, 2003). To broaden the EMS system to ministries and to encompass 'indirect effects' should make a difference if implemented properly. But this remains to be seen. Proper implementation has to do with the organizational location of the EMS function. We have witnessed containment and isolation of environmental management systems in agencies, with one individual or a small group in charge of the procedure without real influence or connection to the strategic level of the organization. A sector environmental analysis, as performed in Chapter 4, can be useful when assessing 'indirect effects' of the ministries. The environmental impacts of the sector can be seen as the 'indirect effects' of the corresponding ministry (Finnveden et al, 2002).

Organizational provisions for EPI

In addition to procedures discussed above, there is an increasing blend of organizational experiments. The government has made several attempts to create crosscutting groups for dealing with sustainable development. At the ministerial level, there was one council set up in the late 1990s and another one planned under the Ministry of Industry, Employment and Communications (that never met!). At the ministry civil servant level, there is a Coordination Unit for Sustainable Development, created in 2004 under the Prime Minister's Office and moved, in 2005, to the new Ministry for Sustainable Development. There is also a unit to promote environmental regulations across government under the same ministry tasked with the mainstreaming of environmental regulations into all sectors. At the agency level, in 1999 the government created the Institute for Ecological Sustainability under the Ministry of Environment. This was replaced in January 2005 by the Sustainability Council, created under the National Board of Housing, Building and Planning (Boverket). All important sector agencies are also represented by their chiefs at the Environmental Objectives Council, housed under SEPA.

It is difficult to discern the impacts of this accumulation of organizations. It is apparent that certain bodies, such as the Environmental Objectives Council or the Coordination Unit for Sustainable Development, do useful and relevant work in coordination and capacity development. On balance, however, the increasing organizational complexity has made EPI a very messy enterprise and our interviews indicate that it also undermines the credibility of the political initiative itself. Since the potential for EPI as a learning enterprise tends to depend on the initiative and acceptance of the economic sectors, this is clearly problematic. New constellations are created as a way to demonstrate political initiative and will, but people ask themselves, where is the will if the efforts are not followed through?

Research and development

Research and development can play key roles for learning and for EPI in several ways. Different types of assessments can play an important role for EPI as discussed above. These assessments will, however, often be dependent upon previous research. For instance, the committees discussed above can be important arenas for learning, but they do not normally have resources to produce new knowledge given their time and budget constraints. They are thus dependent on research that is already available. Therefore an important way of promoting EPI is to support research and researchers that may become important in the policy arena during coming years. This requires foresight and economic resources among organizations that wish to promote EPI. Research and development can also be used to create stakeholders within sectors. An example was noted above concerning bioenergy, where long-term support of R&D has created stakeholders within the energy sectors who will benefit from putting environmental aspects on the agenda. Research about environmental problems and their solutions can also be used to change values and norms among policymakers at different levels, as well as the general public, influencing the acceptance of different types of decisions. This has been notable in the agriculture sector, where the Swedish University of Agricultural Sciences has played an important political role.

Reflections on Concepts and Methodology

EPI, just like sustainable development, has been subject to different interpretations and, sometimes, conflicting perspectives. Analysts have observed a goal conflict in pursuing both horizontal (networking) and vertical (implementing) modes of EPI. In the former, actors are expected to be empowered and motivated to bring about change at the same time as watching over their more narrow interests. In the latter, the government has, through overarching democratic procedures, defined the goals and requested that sectoral agencies and associated actors implement them. As Sabatier (1986) argued in his work on implementation, a policy-learning approach might be considered as a third route – navigating between and building on the perhaps normatively appealing, but unrealistic top–down norm, as well as the power game that sometimes characterizes bottom–up processes. EPI in this study is defined as 'a policy-learning process in which perspectives evolve and sectoral actors reframe their objectives, strategies and decision-making processes towards sustainable development' (Nilsson, 2005). The learning perspective on policy integration has conceptual advantages in that it seems to capture the essence of the intentions behind the political principle and is compatible with the increasingly established network perspective on policymaking. It should be noted that this learning approach contrasts this study with other EPI analyses. In the mainstream conceptions, EPI is often thought of as a matter of giving priority to environmental over other objectives in policy decisions (Lafferty and Hovden, 2003). However, such measures do not in themselves differentiate the causes behind how environmental issues are brought in. This is problematic because long-term implications can be very different if a change in environmental priority results from power shifts rather than from a cumulative learning process.

The methodology in this book has posed challenges, conceptually as well as empirically. Concepts of learning and frames have been called a 'conceptual minefield' (Levy, 1994). What is, for instance, the real difference between technical and instrumental learning, or between frames and value systems? We have not dwelled at any length on this. Instead, we have addressed what we saw as a lack of in-depth empirical work relating to EPI. We have tried to develop a framework for learning and frames that cover the main concepts in a pragmatic way. In doing so, theoretical nuances have been lost. The empirical representation of learning also has its problems; learning-based policy change needs to be distinguished from – and related to – changes due to, for instance, power shifts among interest groups or in parliament. This distinction is not trivial, but was achievable through studying policy change over sufficiently long time periods and through a triangulation of analytical approaches including argumentation and debates, policy outputs and content analysis. This gave a reasonably robust yet dynamic representation of the framing and EPI patterns in the sectors. In the first stage of the analysis a highly aggregated view of EPI was deployed. In this view, we could discern some patterns. Through the contents analysis we were then able to deconstruct the environmental issue somewhat and found patterns that confirmed the findings of the frames analysis, as well as adding insights, for example how growing attention to climate change crowded out the concern for other environmental issues of relevance, such as biological diversity, landscapes and air pollution.

The learning approach assumes that the self-serving incentives of policy officials are not part of the model, although such incentives (in terms of career, financial reward or status) have shown strong explanatory value in other contexts, as discussed in the political economy literature. The assumption is that, relative to most other countries, Swedish policymaking is governed more by an institutional 'logic of appropriateness' than self-interest. That assumption can be made by looking at how Swedish bureaucracy career paths and incentives are created, where there is a strong value tradition of the honourable and loyal civil servant. However, our interviews indicate that the highest political leadership sees things in quite different ways from the administrative parts of the governmental machinery. In their arena, it appears that individual benefits and positioning can become a more important premise. It has not been an objective to trace these differences systematically, but they constitute an important avenue for further understanding policy processes.

The book has given insights on what driving forces and what institutional arrangements can be associated with EPI. However, the exact mechanisms behind this association have been beyond its scope. Organizations change through the learning of their members, in other words through what goes on inside the heads of people that inhabit the policy system. A further unpacking of EPI as learning, building on organizational literature, remains a useful path ahead for future research (Huber, 1991). One approach is to use systematic survey data to capture patterns over time. There is an increasing wealth of survey data on opinions, attitudes and values with regard to different types of issues. This provides an opportunity to create more in-depth research programmes to trace the intricate relations between cognition and action.

A potential problem with the approach taken is that it is based on selective descriptions. In addition, a prior in-depth understanding of the issues and subtleties of policymaking is necessary. It would therefore be difficult for another team to replicate the results with the same approach. However, the logic of the argument, the combination of data sets and displays of evidence give a sufficient basis for our conclusions. Could one imagine a more formalized approach? In our view the complexity of EPI processes argues against a highly formalized model. A quantitative analysis and stronger formalization would lead to precision, but explanations would remain partial and need embedding in larger narratives. Furthermore, quantitative approaches must be thought through very carefully. Often, the quest for something measurable is at the expense of the 'conceptual fit'. Still, more formalized and quantifiable approaches should be further explored; they are needed for governments to trace progress and make cross-country comparisons.

As noted in Chapter 3, only a limited number of published sector environmental analyses include upstream and downstream effects. The development of this approach is thus an important part of this study. It is useful to include impacts caused by decisions made in the sector being studied even if they occur in other sectors, or even in other countries. Some measures that can be taken to reduce environmental impacts within a sector will only move the impacts to another sector or even to another country. If such problem shifting occurs, this cannot be called EPI. Holistic studies like the ones used can also be useful to identify the most important environmental aspects for a sector. As shown, there may be important aspects which are not given the attention they may deserve from an environmental perspective. Such assessments are therefore valuable to make sure that important issues are not forgotten. They can also be used to identify actors that are responsible for environmental impacts and thus where policy interventions can be of interest. One advantage of the assessment methodology is that it relies on data from the System of Economic and Environmental Accounts, which is to some extent standardized and used in many countries. This makes international comparisons possible and interesting. The methodology and data used here can also be used for developing indicators for EPI in a similar way to the current development of indicators for Integrated Product Policy (IPP) (Palm et al, 2006). However, when doing so, it should be acknowledged that environmental impact is a poor proxy measure for EPI, because of the significant time lags between policy interventions and visible improvements in environmental pressures. Furthermore, exogenous driving forces, such as natural variability, macro-economic fluctuations and new technologies, often have a larger influence than policy. Sectors can in some cases make significant progress without significant efforts, and conversely, sectors can fail to make progress despite active policy efforts (Skjaerseth and Wettestad, 2002).

Nevertheless, it is relevant to observe outcomes in an EPI study, as its ultimate aim and raison d'être is to reduce the negative environmental impacts of sector activities. We find that the sector environmental analysis brings information on environmental effects that might not be seen by the sector itself and complements the methods applied in the policy content and frames analyses. Therefore, this study combined several analyses of processes, outputs and outcomes. This, as we know from policy evaluation literature, gives a far more robust albeit more complex picture (Weiss, 1998; Dunn, 2003). We have found the combination of methods from the environmental and social sciences useful when searching for ways to pinpoint EPI and this is, perhaps, the main conceptual and theoretical contribution of our research project.

EPI as Learning – Final Remarks

This book has focused on environmental policy integration as a process of policy learning. The idea that EPI is a process of learning brings certain connotations. In a learning mode, policy processes might be seen as free of conflict and power games, as a rational and informed search for the best solutions keeping in mind all relevant environmental knowledge. We hope that this book has demonstrated that learning entails much more complex processes. In contrast to the rationality norm, it is entangled with interest politics and strategic behaviour. Frame change is partly a subconscious process and partly an intentional strategic process. Actors frame things in new ways in order to advance their cause on the policy agenda. For instance, in the 1990s environmentalists reframed their advocacy from a growth critique to eco-modernization with considerable success. Similarly, proponents of nuclear power have had considerable success in reframing their agenda and position from a low-cost supply, initially, to climate protection. This is a type of political learning (May, 1992) and it means that frames are partly constructed. But it can also trigger learning processes and EPI by constructing argumentative 'bridges' (Van Eeten, 1999). Much of the EPI success in recent years has actually been related to a strategy of building bridges to security, rural development policy and landscapes. In this sense, EPI is a change in perspective that can be triggered by, as well as lead to, a frame change.

Our study suggests that moderate levels of tension and conflict are probably prerequisites for policy learning to occur at all. When sustainable development emerged as a new issue, it was difficult to characterize the policy situation and what actors really represented what interest. Everything was hunky dory – who could be against sustainability? But this is also its weakness: sustainable development and ecological modernization offered 'meta narratives' that were politically attractive, but they hid the inevitable trade-offs and costs involved. Many consider the last five years as two steps back when it comes to sustainable development internationally because environmental issues are meeting stronger resistance and have less political clout than in the 1990s, in Sweden as well as globally. However, we argue that in order to get the policy dilemmas clarified (not only to get clarifications between environmental and economic goals, but also between different environmental goals), underlying conflicts need to be articulated and coalitions formed. This is an inevitable part of the process of learning and integration and a necessary process to advance the principle of sustainability from rhetoric to action.

Neither is learning a power-free process. When promoting EPI as a learning process, it is important to recognize that there is a risk that learning processes benefit those on the inside, reinforcing existing power structures by creating even stronger stakeholders, while other less-resourced participants come out less powerful (Meadowcroft, 2004). Indeed, even under conditions of learning, political decision making finally remains the outcome of power relations between major players in a parliamentary democracy. But learning is a major part of the interest formation of these players. Ultimately, actors may participate in learning processes to demonstrate openmindedness, but still stick to their interests and agenda. In many situations decisions are then made on purely political grounds in spite of well-researched background materials such as environmental impact assessments

(EIAs). There is a range of theories on 'nested games', 'policy windows' and so forth that can help us understand such outcomes (Tsebelis, 1991; Kingdon, 1995).

EPI as a learning process is to a large extent about work behind the scenes. This is perhaps one of the most difficult aspects of the process. Because the learning processes are invisible to public opinion, politicians lack a purely political incentive to engage. For a politician, working for a long-term change behind the scenes is often less rewarding than taking a strong public stand and engaging in a conflict on a particular issue or solution. The learning approach is in this sense not completely in resonance with the nature of democratic politics.

References

Baker, S. (2000) 'The European Union: Integration, competition, growth – and sustainability', in B. Lafferty and J. Meadowcroft (eds) *Implementing Sustainable Development: Strategies and Initiatives in High Consumption Societies*, Oxford University Press, Oxford, UK

Daugbjerg, C. (1998) 'Similar problems, different policies: Policy networks and environmental policy in Danish and Swedish agriculture', in D. Marsh (ed) *Comparing Policy Networks*, Open University Press, Buckingham, UK

Ds (2000:1) 'Kommittéhandboken' ['The committee handbook'], Ministry Publication Series, Regeringskansliet, Stockholm

DsFi (1999:27) 'Rapport från ett ESO-seminarium: Med backspegeln som kompass' ['Report from an ESO seminar: With the rear-view mirror as compass'], Ministry Publication Series, Regeringskansliet, Stockholm

Dunn, W. N. (2003) *Public Policy Analysis – An Introduction*, Prentice-Hall, Englewood Cliffs, US

Eckerberg, K. (2000) 'Sweden: Progression despite recession', in W. Lafferty and J. Meadowcroft (eds) *Implementing Sustainable Development: Strategies and Initiatives in High Consumption Societies*, Oxford University Press, Oxford, UK

EEA (2005a) *Environmental Policy Integration in Europe: State of Play and an Evaluation Framework*, European Environment Agency, Copenhagen

EEA (2005b) *The European Environment: State and Outlook 2005*, European Environment Agency, Copenhagen

Einhorn, E. and Logue, J. (2003) *Modern Welfare States: Scandinavian Politics and Policy in the Global Age*, Praeger, Wetsport, CT, US

Engström, R., Nilsson, M. and Finnveden, G. (2006) 'Characteristics and policy attention of environmental issues in two Swedish sectors – agriculture and energy', submitted manuscript

Finnveden, G., Wadeskog, A., Eriksson, B. N., Johansson, J., Palm, V., Åkerman, J. and Hedberg, L. (2002) *Indirekt miljöpåverkan från försvarssektorn [Indirect Environmental Impacts from the Defence Sector]*, FOI, Stockholm

Forsberg, B., Hansson, H. C., Johansson, C., Areskoug, H., Persson, K. and Järvholm, B. (2005) 'Comparative health impact assessment of local and regional particulate air pollutants in Scandinavia', *Ambio*, vol 34, pp11–19

Hajer, M. (1995) *The Politics of Environmental Discourse: Ecological Modernization and the Policy Process*, Oxford University Press, Oxford

Hey, C. (2002) 'Why does environmental policy integration fail? The case of environmental taxation for heavy goods vehicles', in A. Lenschow (ed) *Environmental Policy Integration: Greening Sectoral Policies in Europe*, Earthscan, London

Holmberg, S. and Weibull, L. (eds.) (2001) *Det våras för politiken: SOM-undersökningen 2001 [Springtime for Politics: The SOM Poll 2001]*, SOM-institutet, Gothenburg, Sweden

Huber, G. P. (1991) 'Organizational learning: The contributing processes and the literatures', *Organizational Science,* vol 2, pp88–115

Huppes, H., de Koning, A., Suh, S., Heijungs, R., van Oers, L., Nielsen, P. and Guinée, J. B. (2006) 'Environmental impacts of consumption in the European Union using detailed input-output analysis', *Journal of Industrial Ecology*, vol 10, in press

Inglehart, R. (1997) *Modernization and Postmodernization: Cultural, Economic and Political Change in 43 Societies*, Princeton University Press, Chichester/Princeton

Isaksson, K. (2001) 'Framtidens trafiksystem? Maktutövning i konflikterna om rummet och miljön i Dennispaketets vägfrågor' ['The traffic system of the future? Power exercise in the spatial and environmental conflicts of the Dennis package], PhD thesis, Linköping University, Linköping, Sweden

Jacob, K. and Volkery, A. (2004) 'Institutions and instruments for government self-regulation: Environmental policy integration in a cross-country perspective', *Journal of Comparative Policy Analysis*, vol 6, pp291–309

Jahn, D. (1998) 'Environmental performance and policy regimes: Explaining variations in 18 OECD countries', *Policy Sciences*, vol 31, pp107–131

Jänicke, M. and Weidner, H. (1997) 'Summary', in M. Jänicke and H. Weidner (eds) *National Environmental Policies: A Comparative Study of Capacity Building*, Springer, Berlin

Jasanoff, S. (1990) *The Fifth Branch: Science Advisors as Policy-Makers*, Harvard University Press, Cambridge, US

Jordan, A. (ed.) (2002) *Environmental Policy in the European Union: Actors, Institutions and Processes*, Earthscan, London

Kingdon, J. W. (1995) *Agendas, Alternatives, and Public Policies*, HarperCollins College Publishers, New York

Lafferty, W. and Hovden, E. (2003) 'Environmental policy integration: Towards an analytical framework', *Environmental Politics*, vol 12, pp1–22

Lafferty, W. and Meadowcroft, J. (eds) (2000) *Implementing Sustainable Development: Strategies and Initiatives in High Consumption Societies*, Oxford University Press, Oxford, UK

Lebow, R. (1984) 'Windows of opportunity: Do states jump through them?' *International Security*, vol 9, pp147–186

Lenschow, A. (2002) 'Greening the European Union: An introduction', in A. Lenschow (ed) *Environmental Policy Integration: Greening Sectoral Policies in Europe*, Earthscan, London

Levy, J. (1994) 'Learning and foreign policy: Sweeping a conceptual minefield', *International Organization*, no 48, pp279–312

Lundqvist, L. (2004) *Sweden and Ecological Governance: Straddling the Fence*, Manchester University Press, Manchester, UK

May, P. (1992) 'Policy learning and failure', *Journal of Public Policy*, vol 12, pp331–354

Meadowcroft, J. (2004) 'Participation and sustainable development: Modes of citizen, community and organisational involvement', in W. Lafferty (ed) *Governance For Sustainable Development*, Edward Elgar, Cheltenham, UK

Miljödepartementet (2003) 'Promemoria: utvärdering av Regeringskansliets miljöledningsarbete i beslutsprocesser m.m.' ['Evaluation of the Government Offices' environmental management work in decision processes etc'], Regeringskansliet, Stockholm

Molander, P., Nilsson, J.-E. and Schick, A. (2002) *Does Anyone Govern? The Relationship between the Government Office and the Agencies in Sweden*, SNS, Stockholm

Nilsson, M. (2005) 'Learning, frames and environmental policy integration: The case of Swedish energy policy', *Environment and Planning C: Government and Policy*, vol 23, pp207–226

Nilsson, M. (2006) 'The role of assessments and institutions for policy learning: A study on Swedish climate and nuclear policy formation', *Policy Sciences*, vol 38, pp225–249

Nilsson, M. and Nilsson, L. J. (2005) 'Towards climate policy integration in the EU: Evolving dilemmas and opportunities', *Climate Policy*, vol 5, pp363–376

Palm, V., Wadeskog, A. and Finnveden, G. (2006) 'Swedish experiences of using environmental accounts data for integrated product policy (IPP) issues', *Journal of Industrial Ecology*, no 10, pp57–72

Prop (1997/98:145) 'Svenska miljömål. Miljöpolitik för ett hållbart Sverige' ['Swedish environmental objectives: Environmental policy for a sustainable Sweden'], Government Bill, Regeringskansliet, Stockholm

Prop (2004/05:150) 'Svenska miljömål – Ett gemensamt uppdrag' ['Swedish environmental objectives – A joint mission'], Government Bill, Regeringskansliet, Stockholm

Rittel, H. and Webber, M. (1973) 'Dilemmas in a general theory of planning', *Policy Sciences*, vol 4, pp155–169

Sabatier, P. (1986) 'Top–down and bottom–up approaches to implementation research: A critical analysis and suggested synthesis', *Journal of Public Policy*, vol 6, pp21–48

Sabatier, P. (1998) 'The advocacy coalition framework: Revisions and relevance for Europe', *Journal of European Public Policy*, vol 5, pp98–130

Scruggs, L. A. (1999) 'Institutions and environmental performance in seventeen Western democracies', *British Journal of Political Science*, vol 29, pp1–31

Skjaerseth, J. B. and Wettestad, J. (2002) 'Understanding the effectiveness of EU environmental policy: How can regime analysis contribute?', *Environmental Politics*, vol 11, pp99–120

Skou Andersen, M. and Liefferink, D. (eds) (1997) *European Environmental Policy: The Pioneers*, Manchester University Press, Manchester, UK

Skr (2005/06:126) 'Strategisk utmaningar – En vidareutveckling av svensk strategi för hållbar utveckling' ['Strategic challenges – Further development of Swedish strategy for sustainable development'], Government Communication, Stockholm, Regeringskansliet

SOU (2005:4) 'Liberalisering, regler och marknader' ['Liberalization, rules and markets'], Government Committee Report, Stockholm, Regeringskansliet

Tsebelis, G. (1991) *Nested Games: Rational Choice in Comparative Politics*, University of California Press, Berkeley, US

Uhrwing, M. (2001) *Tillträde till maktens rum – Intresseorganisationer och miljöpolitiskt beslutsfattande [Access to the Rooms of Power – Interest Organizations and Decision-Making in Environmental Politics]*, Gidlunds förlag, Hedemora, Sweden

Van Eeten, M. (1999) *Dialogues of the Deaf: Defining New Agendas for Environmental Deadlocks*, Eburon, Delft, The Netherlands

Weale, A., Pridham, G., Cini, M., Konstadakopolous, D., Porter, M. and Flynn, B. (2000) *Environmental Governance in Europe*, Oxford University Press, Oxford, UK

Weiss, C. (1998) *Evaluation*, Prentice Hall, Upper Saddle River, US

8

Shaping Institutions for Learning

Måns Nilsson

This chapter tries to distil the most important implications of our work that can serve as lessons for other countries and regions. A fuller summary discussion can be found in Chapter 7.

Are the Lessons Valid Outside Sweden?

It is true that the public and key policymakers in Sweden are relatively conscious about environmental issues. It is also true that the Swedish public administration is relatively effective and capable of achieving its goals. Therefore, Sweden is a most-likely case for EPI to occur: if it does not happen in Sweden, it is unlikely to happen elsewhere. Furthermore, this book is concerned with institutions and governance arrangements, and these need to be adapted to the specific issues they are supposed to address, to the traditions and cultures of the actors engaged in the process, and to the broader social and economic changes that all societies are going through. In essence, it is not possible to prescribe the same institutional arrangements for all countries and for different levels of governance.

What, then, is the relevance of the Swedish experience to another country or to decision-making processes at regional or international levels? The analysis of EPI in Sweden shows that EPI does occur, but only to a certain extent and only under certain circumstances. This has given us a variation in EPI outcomes whose circumstances have been analysed and discussed at length in Chapters 5, 6 and 7. EPI barriers and problems encountered in Sweden are well known internationally and will be encountered elsewhere too. In essence, when we talk about learning and framing we are dealing with social and psychological processes. These are generic phenomena of human nature and social interaction. The principles and mechanisms behind the things that seem to work well in Sweden, and the things that work less well, are therefore broadly relevant.

In other words, most of the things that work well in Sweden should work pretty well in other places as well, and vice versa. However, it should be kept in mind that in Sweden most people in the policy arena have a fair amount of faith in each other's intentions and capabilities in seeing them through. There is a baseline of common norms and rules of conduct between different actors and interests. The same goes for any type of integration, really, whether it is social integration of ethnic minorities into the workforce, or whether it is the technical integration of electricity grids and distribution markets between countries. For integration to occur, it is necessary to establish a minimum set of common norms and rules of conduct in politics, which, surprisingly, is still not the case in many advanced Western democracies. However, it is likely to be only a matter of time (possibly requiring generational shifts in staff) before these things get more engraved throughout societies. Still, in the short term, there are societies and issues where there is no trust between actors, and where there is no basis for discussing a problem together. In these cases you have some groundwork to do before embarking on an EPI strategy of learning. Indeed, this brings to mind the nuclear policy experience in Sweden, where decades of conflict resulted in a basic lack of trust that completely stalled progress. Something similar seems to be occurring today surrounding the issue of agricultural biotechnology, where social conflict is hindering a dialogue on whether the new technology can be adapted to serve environmental and social purposes.

The concrete implication can be the effect, on balance, from moving the mandate and responsibility for environmental impacts into the sector ministries and agencies. When the level of trust and basic norm agreement is high, the positive effects on ownership and accountability can override the negative effects of 'burial' that can happen when mainstream economic concerns are still overriding. In contrast, in places where actors in the economic sectors do not accept that environmental issues matter, the case might well be the opposite: environmental issues get buried. Indeed, looking at mainstream environmental policy literature, burial is normally considered an overwhelming risk. There might therefore be many, both in the environmental movements and among economic interests, that remain unconvinced of the benefits of an integration strategy as outlined in this book. The alternative to EPI, then, becomes the old style of environmental policy: as a discreet area that works like a regulator and watchdog. But even if this has been successful in the past, it can only take us so far, and not as far as EPI, which looks towards a truly sustainable development perspective engraved in the 'minds' of the economic sectors.

Summary Comparison of the Sectors and their EPI

Sweden is seen as a front-runner in environmental policy globally. The institutional as well as political arrangements appear beneficial to EPI. The sectoral policies are nowadays, by and large, considered environmental policies. But what is the EPI situation really like?

Although our two case study sectors are both home to some of today's major environmental problems, and also subject to intensive integration efforts over the last 10–20 years, the two sectors display marked differences in several dimensions. Chapter 5 showed that both sectors have reframed over the last 15–20 years as a result of major contextual changes, but that environmental objectives have become only partially integrated, and in different ways in the two sectors. In agriculture,

EPI has occurred broadly but primarily in a parallel policy track and through the development of an 'agriculture-for-sustainability' frame. In energy, environment has become mainstreamed into the dominant 'energy-as-market' frame. However, the flip side of this more mainstream acceptance is a narrower environmental framing, limited to climate change in energy as compared to a more systematic view of the NEQOs in agriculture. We can therefore talk about a deep but narrow integration in energy against a broader but shallow integration in agriculture.

These ways in which environmental issues become part of the sector framing link to the overall policy context of the sectors. Agricultural policy is in principle determined at the EU level, whereas the decision-making competency over energy policy is still national. However, in practice Europe's influence is probably equally strong in the two sectors today: agricultural policy programmes are tailored nationally, and the EU's internal market and environmental policies of the EU set strong boundary conditions on energy policy. Still, the agricultural sector remains heavily regulated and under state control empowered by the EU's CAP, whereas most parts of the energy sector are deregulated. This means there are still strong opportunities to create direct environmental support programmes – in other words direct subsidies that are well defined and can be disjointed from other sectoral measures. In the energy sector, environmental incentives must be created within market mechanisms, such as certificate systems, which impact on the whole energy market, for instance by driving up prices and profits for the major electricity companies. This appears to be one important reason for the differential EPI patterns observed: the energy-market context forces a deeper integration structure by forcing the instruments to work across the whole market. In agriculture, EPI has so far been contained to the support programme.

Another aspect is the sector configuration of actors. The key actors and productive capacity in agriculture are the many small private farmers and their suppliers and distributors, organized through movements that work from the grass roots up to the national level. The sector is, to a large extent, still dominated by the old corporatist network. In energy, actor constellations have changed dramatically. As the sector restructured there was a strong influx of new actors, perspectives and ideas entering the mainstream of energy policymaking. Still, today a few very large actors (three major electricity companies) dominate the sector once again. Although the explanation is weaker, the relatively small set of actors in energy might relate to the relatively uni-dimensional EPI. In essence, the interest configuration within the sector is less diverse. The wider range of actors in agriculture provides for a broader coverage of environmental issues in the sector, but as we saw above, the policy context provides for containment of these issues rather than forcing them into the dominant framing.

Thus the broader context and the way the sector looks matter a great deal for whether and how EPI can occur. But within those constraints, this book has also looked more closely at institutions that might contribute to making or breaking EPI. The different rules of the game within different levels of the government machinery affect actors' ability to share ideas and engage in long-term thinking across various instances. Below are outlined four principles followed by ten institutional 'tips' of broad relevance when devising a strategy for EPI.

A New Institutional Strategy for Learning?

The usual off-the-shelf prescriptions for EPI are well known – typically including things such as regulatory provisions for inclusive decision-making procedures, setting up coordinating bodies like councils or secretariats, capacity-building and awareness-raising programmes, integrated economic-ecological accounts and indicator systems, impact assessments, and follow-up and evaluation procedures. In our experience, these are all valid measures that *could* have a role contributing to EPI. But whether they do or not depends on other factors. For EPI to happen, their institutionalization is necessary but not sufficient. If we accept the learning notion dimension as a key to EPI, finally we have to dig deeper and ask ourselves the bigger questions: Where and how do ideas develop and disperse? Where and how do they seep into the mainstream? What does a forum for mutual understanding and constructive tension, creativity and new knowledge assimilation look like? The devil is probably in the detail, and the details must, as discussed earlier, be worked out according to the specific issues, contexts and actors. However, if we want to move in this direction, some institutional principles should be at the forefront of the design.

First, an institutional strategy must recognize that actors take a risk when involving themselves in testing innovative ideas and learning. Their participation is therefore driven by the perception that the benefits from cooperating are greater than the risks of losing ground in the policy arena in one dimension or another. This requires a balance in how close the forum is to the political process; if we are really close to decision making, the perceived risks become too large. If we are detached from political processes, the perceived benefits become too small.

Second, an institutional strategy must focus on development of trust and joint-problem perceptions. This is at the core of the EPI process: people first need to recognize that sustainability is a complex problem that requires a complex response, and a 'wicked' one since responses tend to create new problems elsewhere. Under the right conditions, through efforts where participants start to entangle these issues and share the dilemmas and trade-offs that emerge, joint-problem perceptions across actors and trust in each other's intentions can start to surface. Eventually, this can propel into breakthroughs in understanding between people from traditionally opposing organizations.

Third, an institutional strategy must seek out ways to align the reframing with the dominant perceptions and behaviours in the existing structures. Our evidence suggests that actors are indeed willing and able to reframe and integrate environmental values, but that they will only be supported in political decision making if they somehow align, or create alliances, with important shorter-term interests within the sector. Luckily, the reframing process eventually also affects how actors understand their short-term interests.

And fourth, strong political leadership is essential to back up the institutional strategy. Without support from the higher spheres of central government, efforts to ensure EPI will compete with a range of other political and administrative priorities and risk getting lost in complex decision-making structures. Our studies have shown that EPI has been most successful in times of political collaboration at high levels promoting environmental concerns.

Ten Tips on Shaping Institutions for EPI

What, then, is needed to embark on a learning strategy to EPI? This study has pointed us towards four important framework conditions:

- *trust:* different stakeholders within and outside government must have trust in the processes and governance;
- *ownership:* sectoral actors need to take environmental issues to heart, which they will do only if they are competent to address them and held accountable for delivering results;
- *capacity:* the capacities to engage in knowledge assimilation, interpretation, strategic thinking and interactions with different stakeholders are all required – activities that take time and require highly qualified officials; and
- *knowledge:* integrating environmental issues depends on how well sectors understand the environmental ramifications of their strategies and activities.

Below are ten institutional tips specifically addressing these conditions and helping to shape the institutional landscape to enable effective implementation of EPI.

1 **Firmly institutionalize high-quality impact assessment.** Good environmental decision making requires that actors understand the broader consequences of their positions and choices. Obtaining this understanding cannot be subject to a fluctuating political will but needs to be mandatory and widely accepted. Such institutionalization is presently more advanced at the level of the European Commission and in many other European countries than in Sweden. However, the quality still tends to be poor in most places, undermining its credibility and usefulness. Therefore, impact assessments need to be coupled with strong quality controls, opportunities for public scrutiny and independent review procedures.

2 **Make ministries the central points in EPI efforts.** Ministries have the most central role in policy formation and must be the key agents for EPI. The tendency to devolve EPI efforts to agencies is problematic. However, there is a knowledge deficit at the strategic levels. This deficit is not primarily coupled to capacity constraints but more to the incentives and informal structures shaping ministry operations. EPI advocates must better understand these constraints and recognize that they are different at different levels of the process.

3 **Enable openness of central government for dialogue across ministerial and sectoral borders.** The tradition that each ministry only looks after a narrow set of interests induces strategic behaviour and limits communication and coordination. Until recently, ministries did not allow their agencies to collaborate with other ministries' agencies. When this changed, important patterns of EPI emerged in agency processes. This is a matter of signals from above but also about institutional capacity development to enable interactions at multiple levels.

4 **Reallocate competencies and mandates from one ministry to another.** The reallocation processes associated with reorganizations and the sector responsibility/NEQO schemes have shown promising results, although they also entail significant costs. When the environmental bureaucracy cedes some competency in resolving environmental issues to sectors, the sectoral borders between ministries and agencies become more passable.

5 **Apply more *ex post* evaluation.** The focus on developing *ex ante* impact assessments to provide decision support is necessary, but at least as much effort must be given to learning from the past. Often this will be more useful. Indeed a recent surge in the interest in *ex post* policy evaluation appears to be a positive boost both for EPI and for policy learning more generally. Rigorous feedback procedures are needed to ensure that all relevant actors become informed.

6 **Innovate in public auditing systems for environmental policy.** Systems to evaluate environmental policy performance are an as yet untapped opportunity for public scrutiny, accountability and bounding of politics. Rather than pushing their mandate to include environmental concerns, it is worth considering separate auditing bodies for environmental or sustainability affairs. The UK Environmental Audit Committee and the New Zealand Parliamentary Commissioner for the Environment represent two interesting models.

7 **Develop new environmental-analytical tools for EPI.** More comprehensive tools for assessment are needed, across environmental issues but also across sectoral borders. The life-cycle approach to analysing the sector activities applied in this study provided new perspectives on the environmental aspects of sector activities and what remedies can be taken. However, because of the fuzziness of the alternatives and their properties, the lack of well-stated preferences of various actors, and imprecise data, providing direct 'answers' based on, for instance, multicriteria analysis or cost–benefit analysis is probably not very effective. Such tools have a role to play, but more as iterative assessments for learning about values.

8 **Nurture actor interests and 'issue champions'.** Given the attention span of the political system, even though there might be comprehensive environment and sustainability agendas established, it is usually unable to maintain a comprehensive view of sustainability from agenda setting to decision making. Our study shows that issues need champions within the sector and preferably strategic coalitions between different champions. Active formation of actor networks is needed to create the momentum.

9 **Build trust by acknowledging complexities and trade-offs.** In the Swedish example, the energy sector needs not only to look at climate and resources, but also to pay more attention to forests resources, eutrophication, acidification and landscape issues related to energy systems. The agricultural sector needs to consider broader issues like their contribution to dealing with climate change and marine areas. The inevitable dilemmas and trade-offs between these objectives and with economic and social objectives will lead to tensions between interests. But under the right conditions of trust, tensions can be deployed creatively for learning purposes, instead of fuelling conflict.

10 **Create conceptual learning platforms.** Processes where learning has occurred can be used as models to build on. In this study, we learned that, for policy learning to occur: a) there needs to be moderate (but not too strong) political pressure and interest; b) the process should be neither totally unconnected from, nor directly instrumental to, the decision process; and c) there must be room for interactions that are partly or fully closed from the public vision. Highly visible processes easily create position wars and formalistic behaviour that are not conducive to learning.

Index